Neuroimmune System

The book explores the intricate connections between the nervous and immune systems in the context of neurodegenerative disorders, offering a comprehensive overview of the bidirectional communication between these systems and their implications for disease progression and therapeutic interventions. The book aims to understand the recent developments in the field of neuroimmune communication.

Key Features:

- In-depth analysis of immunological biomarkers and therapeutic targets for neurodegenerative disorders.
- Exploration of the role of glial cells in neuroimmune interactions and their dual nature in disease pathology.
- Examination of short- and long-range interactions between the central nervous system (CNS) and immune system, including hormonal axis and cytokine milieu.
- Discussion on the impact of microbiota on neural networks and immune responses, highlighting the gut-brain axis and potential therapeutic avenues.
- Utilization of Drosophila model systems to study neurodegeneration, providing insights into disease mechanisms and potential treatment strategies.
- Coverage of various neurodegenerative disorders, including Alzheimer's, Parkinson's, polyglutamine-related diseases, and tauopathy, with a focus on understanding disease progression and identifying therapeutic targets.

The book stands out for its interdisciplinary approach, bridging the fields of neuroscience, immunology, and microbiology to provide a holistic understanding of neurodegenerative diseases. It offers valuable insights into the complex interplay between the nervous and immune systems, shedding light on novel therapeutic avenues for treating these debilitating conditions. Suitable for researchers, clinicians, and students in the fields of neurology, immunology, and microbiology, this book serves as an essential resource for understanding the pathogenesis and treatment of neurodegenerative disorders.

Neuroimmune System

Edited by
Arpit Kumar Pradhan
Ashwin Kumar Jainarayanan
Shyamasree Ghosh

CRC Press
Taylor & Francis Group
Boca Raton New York London

CRC Press is an imprint of the
Taylor & Francis Group, an informa business

First edition published 2025
by CRC Press
2385 NW Executive Center Drive, Suite 320, Boca Raton FL 33431

and by CRC Press
4 Park Square, Milton Park, Abingdon, Oxon, OX14 4RN

CRC Press is an imprint of Taylor & Francis Group, LLC

ISBN: 978-1-032-28140-7 (hbk)
ISBN: 978-1-032-28143-8 (pbk)
ISBN: 978-1-003-29547-1 (ebk)

DOI: 10.1201/9781003295471

Typeset in Times
by SPi Technologies India Pvt Ltd (Straive)

Contents

Preface

Neuroimmune communication is perhaps the most intriguing domain of biology, of the day, the study of which has immense importance in understanding the intricate details of communication between the nervous system and immune cells during the process of development, homeostasis and in pathogenic and infectious disease conditions.

Immunology and neuroscience are two complex domains and the immune responses in the central nervous system (CNS) and those in the periphery reveal differences. Anatomical barriers and neuroglial cells protect the brain parenchyma from potentially harmful damage mediated by immune cells, but immunological control in the CNS enables the ability to confront and contain neurotropic pathogens, deregulation of which leads to neuroinflammatory diseases. The peripheral immune system and CNS communicate by molecular signals.

This book highlights the biology of the Neuroimmune system, highlighting the neuroimmune communications in healthy individuals and their deregulation associated with disorders of the neuroimmune system. It highlights the neuroimmune communication system scenario in neurodegenerative disorders. Discussing the scenario and the state of art in humans, and animal models systems of Drosophila, we discuss the neuroimmune system in Canines associated with neuromuscular disorders. We also discuss the neuroimmune system and its biological function in Infection. The gut microbiota is known to modulate immune function of cells and finds importance as a regulator in brain development, function, and behavior. We discussed the role of gut microbiota in neuroimmune regulations with insights into gut neuropeptides and gut hormones.

Arpit, Ashwin and Shyamasree

Editor Biographies

Dr. Arpit Kumar Pradhan Department of Psychology, University of Michigan, Ann Arbor, MI 48109, USA

Dr. Arpit Kumar Pradhan is currently a postdoctoral research fellow at the University of Michigan working with Professor Ada Eban-Rothschild to study the role of sleep in memory consolidation. He has expertise in the application of behavioral neuroscience, electrophysiology, and advanced imaging techniques to study neurodegeneration and develop potential therapeutic interventions. He completed his doctoral degree (PhD) from the Graduate School of Systemic Neurosciences at the Ludwig Maximilian University of Munich under the guidance of Professor Gerhard Rammes. During his doctoral program, Dr. Pradhan studied the impact of translocator protein (TSPO) activation on amyloid beta-induced synaptotoxic effects and showed that TSPO ligands such as XBD173 via delta subunit-containing $GABA_A$ receptors ameliorates the pathophysiology in Alzheimer's disease. He has completed his master's from NISER Bhubaneswar, India, and has spent substantial research time working in the field of neurodegeneration at NUS Singapore, IIT Bombay, JNCASR Bangalore, and TIFR Mumbai. His research interests include studying sleep and memory consolidation, neuro-glia interactions and their potential effect on behavior, and neurodegeneration.

Dr. Ashwin Kumar Jainarayanan Interdisciplinary Bioscience Doctoral Training Program and Exeter College, University of Oxford, Oxford, UK

Kennedy Institute of Rheumatology, Nuffield Department of Orthopedics, Rheumatology and Musculoskeletal Sciences, The University of Oxford, Oxford, UK

Dr. Ashwin Kumar Jainarayanan is an interdisciplinary researcher with a passion for translational medicine. They specialize in computational biology, applying next-generation sequencing technologies to dissect immunosuppressive mechanisms in the cancer microenvironment. Dr. Jainarayanan completed their DPhil/PhD in Interdisciplinary Bioscience (BBSRC DTP) under the co-supervision of Professor Michael Dustin and Professor Matthew Wood. Their doctoral research focused on characterizing extracellular particles released into the immunological synapse, a nano-scale gap between T cells and antigen-presenting cells. Prior to their doctoral studies, Dr. Jainarayanan graduated with a BS-MS dual degree from the Indian Institute of Science Education and Research (IISER), Mohali, India. Dr. Jainarayanan's research interests center around the intersection of immunology, synthetic biology, and neuroscience, with a particular emphasis on addressing critical challenges in cellular diagnostics and targeted molecular therapies.

Dr. Shyamasree Ghosh Scientific Officer (F) at NISER Bhubaneswar, India

Dr. Shyamasree Ghosh has several years in service in the post of Scientific Officer in NISER Bhubaneswar, India. She has graduated from the prestigious Presidency College, Kolkata, with Masters in Biotechnology from Calcutta University, ranking second in the University. She did her PhD in Cancer Biology from the Indian Institute of Chemical Biology (IICB), Kolkata, with Post from the Indian Association of the Cultivation of Sciences (IACS, Kolkata). Dr. Ghosh has extensively worked with the immune system and interdisciplinary domains of stem cells, carbohydrate biology and computational biology, she has authored several books, and published in interdisciplinary domains.

Contributors

Suchandra Chowdhury
Post Graduate Department of Zoology
Bethune College
Kolkata, India

Mirudula Elanchezhian
Department of Biological Sciences
Indian Institute of Science Education and Research
Mohali, India

Shyamasree Ghosh
School of Biological Sciences
National Institute of Science Education and Research (NISER)
Bhubaneswar, an OCC of Homi Bhabha National Institute (HBNI)
Odisha, India

Ashwin Kumar Jainarayanan
Kennedy Institute of Rheumatology
Nuffield Department of Orthopedics, Rheumatology and Musculoskeletal Sciences
The University of Oxford
Oxford, United Kingdom
and
Interdisciplinary Bioscience Doctoral Training Program and Exeter College
University of Oxford
Oxford, United Kingdom

Ranjeet Singh Mahla
Kennedy Institute of Rheumatology
Nuffield Department of Orthopedics, Rheumatology and Musculoskeletal Sciences
The University of Oxford
Oxford, United Kingdom

Sanjima Pal
Research Institute-McGill University Health Centre (RI-MUHC)
Montreal, Quebec, Canada

Arpit Kumar Pradhan
Department of Psychology
University of Michigan
Ann Arbor, United States
and
Klinik für Anaesthesiologie und Intensivmedizin der Technischen Universität München
Klinikum rechts der Isar
Munich, Germany
and
Graduate School of Systemic Neurosciences
Ludwig-Maximilians-Universität München
Martinsried, Germany

1 The Biology of Neuroimmune Communication

Arpit Kumar Pradhan
University of Michigan, Ann Arbor, MI, USA

Mirudula Elanchezhian
Indian Institute of Science Education and Research, Mohali, India

1.1 INTRODUCTION

Over the past decade, our understanding of the crosstalk between the nervous and immune systems has significantly advanced. From understanding the neuroimmune interaction during physiological to studying the disturbance in pathological states we have come far ahead. Our understanding of the neuroimmune interaction is being explored to develop sophisticated multimodal therapeutic paradigms for different classes of neurodisorders. The immune responses in the brain are unique and differ widely from the periphery. In this chapter, we will look into the early development of neuroimmune communications, explore the crucial communicators/messengers (neurotransmitters), and understand in detail the crosstalk between the nervous and the immune systems.

1.2 THE IMMUNE SYSTEM OF THE BRAIN

1.2.1 Early Development of Neuroimmune Communications

The development of neuroimmune crosstalk starts early in the process and is regulated by several key players that shape and modulate the functionality of the system as a whole. The crosstalk of the gut-brain axis starts early during pregnancy and could have an impact on cognitive development. Maternal gut bacteria regulate the maternal immune cells to produce effector molecules, which serve as essentials in fetal brain development (Romano-Keeler and Weitkamp 2015). One such study suggests that the mice fed with a high-fat diet induce changes in the gut microbiota environment, thereby affecting the immune cell function of pregnant mice (Romano-Keeler and Weitkamp 2015; Di Gesù et al. 2022). The offspring of these mice show social deficits and are deficient in the neuromodulator oxytocin in the Ventral tegmental

DOI: 10.1201/9781003295471-1

area. However, upon administration of *L. reutori*, the social alterations as well as decreased oxytocin levels were restored, suggesting that the maternal gut-brain axis impacts the development of offspring and shapes the gut-brain axis of the offspring (Buffington et al. 2016).

One of the recent studies has highlighted that the composition of gut microbiota could be altered during pregnancy (Koren et al. 2012). For instance, the maternal microbiota and immune cells support the development of the fetus during pregnancy (Huh and Veiga-Fernandes 2020). Alterations in the maternal gut bacteria composition in late pregnancy with low doses of antibiotics have shown to have significant behavioral and cognitive alterations in the offspring (Leclercq et al. 2017). Similarly, lack of maternal microbiota affects the development of the male embryonic microglia (Thion et al. 2018). In another study, exposure to elevated levels of maternal immune activation affects the cognition and behavior of the offspring. Pregnant mice that are exposed to viral infection or are injected with synthetic double-stranded RNA (mimicking viral infections) produce offspring with behavioral abnormalities and neurodevelopmental disorders (Shi et al. 2003; Malkova et al. 2012). All this evidence points toward the formation of systemic crosstalk early in the process of development and highlights the importance of the maternal gut-brain axis in shaping the development of the offspring.

1.2.1.1 Story of the "Crosstalk" and How It Impacts Brain Development

1.2.1.1.1 Glial Cells- Innate Immunity

Glial cells hold together the framework of the brain and are essential for its development. Astrocytes, Microglia, Oligodendrocytes, and Satellite Cells are some of the different types of Glial cells that play substantial roles in reshaping and modulating the neuronal circuitry and maintaining efficient functionality of the brain. Astrocytes, the star-shaped cells with several processes, form the most common type of glial cells in the brain and are essential in regulating synapse connections, recycling of the excess neurotransmitters, and development of early immune response in the brain (Wei and Morrison 2019). Their involvement in the regulation of the complement cascade is essential in synaptic pruning. Similarly, microglia, which are much smaller as compared to astrocytes, are derived from monocytes and are responsible for synapse refinement and modulation of brain development (Matejuk and Ransohoff 2020). Oligodendrocytes play a key role in the signal conductance in the brain and are the myelinating glia of the nervous system (Domingues et al. 2016). During nerve injury and neurodisorders, they communicate with microglia and regulate nerve signaling. Satellite cells that are found in the peripheral nervous system (PNS) regulate the chemical environment and are involved in the chronic pain mechanism (Ohara et al. 2009). Chemokine and cytokines are the signaling molecules that are released in the brain by glial cells and other cell types which help facilitate communication within the brain (Banisadr et al. 2005).

1.2.1.1.2 Adaptive Immunity in the Brain

Early in the process of brain development, a small subset of T helper cells migrate and reside in the meninges of the brain. During the course of development, they learn to recognize different brain components such as oligodendrocytes (Spichak, Dinan, and Cryan 2019). The interaction of both innate and adaptive immunity is essential in developing the functionality of the brain (Spichak, Dinan, and Cryan 2019).

1.2.1.1.3 Gut Microbiota

The gut microbiome composition as we have seen earlier plays an essential role in shaping the offspring's gut-brain axis and supports the effective development of the brain. The highly diverse gut microbiota environment plays an important role in shaping the immune system as well as producing a barricade to stop the bacteria as well as their products from entering the bloodstream (Wu and Wu 2012).

1.2.1.1.4 Immune System Development During the Early Phase

Microbial colonization occurs during birth, and the composition of microbiota depends on both genetic as well as environmental parameters. The stabilization of the microbiome environment occurs in parallel with the development of the brain (Spichak, Dinan, and Cryan 2019). In the first stage, they also help in the formation of the mucus layer in the gut and regulate the development of the adaptive immune system. This thick mucus layer prevents the infiltration of bacteria or their products in the bloodstream (Spichak, Dinan, and Cryan 2019). In one of the recent studies, short-chain fatty acid produced by the microbiota has been implicated in the maturation of microglia in early life and their functioning in a sex-specific manner (Spichak, Dinan, and Cryan 2019). Among the environmental factors, stress has been implicated in affecting the microbiota composition. Maternal factors including stress mediate fetal inflammation via interleukin 6 and regulate the response of WBCs to bacterial metabolites. Alterations of immune cells in early life negatively impact brain development (Spichak, Dinan, and Cryan 2019).

1.2.1.1.5 Neuronal Connection Formations and Synaptic Pruning

Brain development occurs by reshaping and formation of new connections. This developmental process is more like a learning experience in which new connections are made and old, unnecessary, connections are removed. Both astrocytes and microglia are two major players that shape and structure these synaptic connections by a process called synaptic pruning. Both these glial cells crosstalk with each other in the form of different cytokines. C1q of the complement cascade is one such essential tag, which is also referred to as the "eat-me" signal and plays an essential role as a signaling molecule in the pruning of neurons (Pradhan et al. 2023). Synaptic pruning in the brain occurs via both C1q-dependent and independent pathways. Subsequently, microglia, upon identification of the tag, release inflammatory cytokines to initiate the engulfment of the synapse through phagocytosis. Conversely, in the C1q-independent pathway, astrocytes recognize an alternate tag through the involvement of multiple EGF-like-domains 10 (MEGF10) and MER tyrosine kinase (MERTK) receptors, resulting in the destruction of the synapse via phagocytosis. (Jung and Chung 2018). Both pruning and the formation of new synapses help develop the brain. This pruning by glial cells plays an essential role as a "double edged sword" in several neurodisorders (which are discussed in the next chapter Figure 1.1).

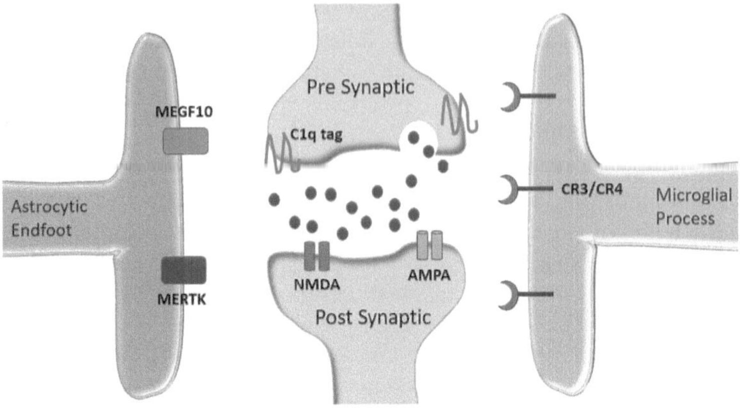

FIGURE 1.1 Synaptic pruning via astrocytes and microglia. C1q eat-me tags are present on the synapses destined to be engulfed by the astrocytes and the microglia. The end processes of astrocytes, which carry receptors such as MEGF10 and MERTK, can identify the C1q tags on the synapses.

1.2.2 Neuroimmune Circuitry and Inter-Organ Development

Interaction between neuronal and immune cells occurs at peripheral sites, and this communication with the central site regulates the function and development of organs. In this part, we will learn about the communication between neuronal and immune cells at different sites, and how this communication impacts the functionality of the particular organ.

1.2.2.1 Neuroimmune Communication in Skin

The diverse plethora of immune cells present in the skin affects the function and regeneration of skin neurons, and this interaction is known to be affected in several disorders (Huh and Veiga-Fernandes 2020). For instance, cytokine IL4 can induce chronic itching via neuronal signaling through the IL-4 receptor-α (IL-4Rα) and Janus kinase 1 (JAK1) pathway (Oetjen et al. 2017). Similarly, mast cells contribute to pain and itch via interaction with sensory neurons (Gupta and Harvima 2018). House dust mites lead to the development of allergic skin inflammation via the nociceptor neuron-mast cell interaction, which leads to the production of Substance P (neurotransmitter) (Serhan et al. 2019). The interaction occurs in the other direction as well, in which neuronal cells in the skin regulate the immune cells. The $TRPV1^+$ nociceptor sensory neurons release calcitonin gene-related peptide (CGRP), a vasodilator neuropeptide, which suppresses pulmonary neutrophil recruitment and activation during *S. aureus* lung infection (Pinho-Ribeiro et al. 2018). Similarly, in cases of microbial infection, the dendritic cells interact with skin nociceptor neurons to regulate γδ T cells through IL23 to induce an inflammatory response (Riol-Blanco et al. 2014; Kashem et al. 2015). The activation of γδ T cells increases the recruitment of neutrophils and monocytes to the local affected site (Riol-Blanco et al. 2014). All these evidences point out to harmonic neuro-immune interaction, which facilitates smooth functioning and a controlled balance in the skin.

1.2.2.2 Neuroimmune Communication in Lungs

Both inflammatory and neuronal cells have been found in the airways, and several studies show this synergistic cross-talk is essential to regulate allergic reactions (Huh and Veiga-Fernandes 2020). In case of exposure to allergen, mast cells and eosinophils are seen to cluster around neuronal cells in the lungs (Nilsson et al. 1990; Jacoby, Costello, and Fryer 2001; Voisin, Bouvier, and Chiu 2017). The intercellular adhesion molecule 1 (ICAM1) released by pulmonary neurons regulates eosinophil recruitment (Huh and Veiga-Fernandes 2020). The interaction between eosinophil and cholinergic neurons in the airways releases granule protein, which influences the nerve cell function via an inhibitory neuronal M2 (Muscarinic) receptor (Costello et al. 1997). The vasoactive intestinal peptide (VIP) released from the neuronal cells can regulate the production of histamine, prostaglandin D2, IL6, and tumor necrosis factor (TNF), which induces the allergic type 2 immune response (Ansel et al. 1993; Kulka et al. 2008; Cyphert et al. 2009). Pulmonary neuroendocrine cells (PNECs), a specialized group of cells located closer to the Group 2 innate lymphoid cells (ILC2), and peptidergic neurons are known to regulate goblet cells via the release of γ-amino butyric acid (GABA) inhibitory neurotransmitter (Barrios et al. 2017). The crosstalk between PNEC and ILC2 cells is known to regulate the allergy-mediated response. In cases of bacterial pneumonia, the pulmonary nociceptor neurons inhibit the neutrophil and recruitment of γδ T cells (Huh and Veiga-Fernandes 2020).

1.2.2.3 Neuroimmune Interaction in Intestine

The gastrointestinal tract receives innervation from both sympathetic and parasympathetic neurons as well as enteric neurons (Godinho-Silva, Cardoso, and Veiga-Fernandes 2019). Bidirectional neuroimmune communication regulates enteric homeostasis, gut health, and peristalsis. The interaction of mast cells and enteric neurons is one such example (Huh and Veiga-Fernandes 2020). Mast cell-derived histamine and tryptase regulate myenteric and submucosal neurons (Tamura and Wood 1992; Steinhoff et al. 2000; Reed et al. 2003). Similarly, in the other direction, peptidergic neurons produce CGRP, Substance P, and VIP, which regulate the mast cells (Shanahan et al. 1985; Kulka et al. 2008). Just like airways, ILCs are commonly located in the mucosal enteric barrier and interact with the neuronal cells. ILC3 via neurotrophic factors belonging to the glial-derived neurotrophic factor (GDNF) family ligands interact with enteric glia (Huh and Veiga-Fernandes 2020).

1.2.2.4 Neuroimmune Interactions in the Liver

The liver predominantly receives efferent vagus nerve preganglionic fibers from the left dorsomedial nucleus (DMN) (Berthoud and Neuhuber 2000). A smaller subset of the hepatic innervations also comes from the right DMN as well as the left anterior nucleus accumbens (NA) (Chan et al. 2023). These innervations are known to play a critical role in hepatic homeostasis, such as gluconeogenesis (Yi et al. 2010). Electrical stimulation of the vagus nerve has been shown to down-regulate the LPS-induced production of TNF in the liver via the release of acetylcholine (Metz and Pavlov 2018).

The neuroimmune crosstalk is essential in liver regeneration. The regeneration of the liver is facilitated by several guidance cues. Netrin 1 is one such neuroimmune guidance cue whose role has been highlighted in liver regeneration. This guidance protein is functional both as a chemoattractant and a chemorepulsant during axonal migration via different subsets of receptors (Ziegon and Schlegel 2021). Netrin 1 regulates the generation of growth factors that contribute to liver regeneration (Ziegon and Schlegel 2021). This highlights the importance of neuroimmune communication in liver regeneration as well as its importance in both chronic and acute disorders.

1.2.3 Neuroinflammation and the Cytokine Network

Cytokines, which form a major part of cell-cell communication, are known for their pleiotropic nature (Becher, Spath, and Goverman 2017). This cell-specific nature of cytokine adds to the complexity of its working mechanism. Chemokines, interferons (IFN), interleukins (IL), lymphokines, and tumor necrosis factor (TNF) form the major class of cytokines (Becher, Spath, and Goverman 2017). In addition to the cell-specific pleiotropic nature of the cytokines, their mode of action and effect vastly differ from tissue to tissue. In this particular section, we will deal with specific cytokine networks in the CNS, how they are a part of homeostasis, and their effects on neural injury and degenerative diseases.

The blood-brain barrier in the CNS restricts the migration of leukocytes into the CNS. During the embryonic developmental phase, the formation of CNS is tightly regulated via cell-cell interactions, and cytokines play a crucial role in mediating the complex organizational framework of mammalian CNS (Becher, Spath, and Goverman 2017). Previously, genetic loss of cytokine receptors has been associated with malfunctions in CNS development (Hatta et al. 2002; Boulanger 2009). Both TGFβ and IL-6 proteins play a crucial role in the neuronal development (Becher, Spath, and Goverman 2017). For instance, the bone morphogenic proteins (BMPs), which belong to the TGFβ class of proteins, regulate the neural differentiation process and the IL-6 class ligands mediate the determination of cell fate and the differentiation patterns in the neuronal and glial cells (Becher, Spath, and Goverman 2017). In addition to the development of neuronal cell types, cytokines have also been pivotal in regulating the growth of glial cells. The role of cytokines in regulating the growth of glial cells is exemplified by IL-34, a key cytokine essential for maintaining the growth and homeostasis of CNS-resident microglia (Becher, Spath, and Goverman 2017). Microglial survival and development are both affected by IL-34, which is produced primarily by neurons in specific brain regions, such as the cortex, anterior olfactory nucleus, and hippocampus (Greter et al. 2012; Mizuno et al. 2011; Y. Wang et al. 2012). Leveraging the role of IL34 in regulating microglial cell homeostasis, a previous study has shown the association of IL34 with improvements in associative learning and cognitive phenotype in an Alzheimer's rodent model (Mizuno et al. 2011). Given the association of reactive gliosis with the pathophysiology of Alzheimer's disease, the cytokine interaction network might play a crucial role in mediating the pathological phenotype of neurodisorders.

1.2.4 Key Players (Neurotransmitters) in the Neuroimmune Network

The neuroimmune crosstalk is mediated by various molecules, including neuropeptides, neurotransmitters, chemokines, and cytokines. Among several neurotransmitters, the most crucial ones are dopamine, glutamate, serotonin, and γ-aminobutyric acid (GABA). Dopamine belongs to the catecholaminergic and phenethylamine families. L-tyrosine is the precursor molecule of dopamine and is synthesized by tyrosine hydroxylase in the substantia nigra pars compacta and the ventral tegmental area (Hodo et al. 2020). The primary function of dopamine in the brain includes motivation, reward, pleasure, and arousal, and is also involved in motor regulation. In the peripheral system, dopamine acts as an exocrine or as a paracrine messenger (Hodo et al. 2020). Additionally, dopamine is also synthesized and stored by immune cells such as activated T_H, and T_{reg} (Cosentino et al. 2000, 2007; Nakano et al. 2009). The dopaminergic receptors are G-Protein coupled receptors (GPCRs), which are metabotropic receptors that modulate downstream pathways via a second messenger (Hodo et al. 2020). These dopaminergic receptors are usually categorized into 5 subtypes (D_1-D_5). The dopaminergic signaling in the immune cells is known to mediate the activation and stimulation of several cytokines. For instance, while D_3 activation modulates TNFα and IFNγ, D_2 activation mediates human T cell secretion of IL-10, and D_5 regulates TNFα and IL-10 secretions (Besser, Ganor, and Levite 2005).

5-hydroxytryptamine (5-HT) or serotonin is synthesized primarily in the Raphe nucleus in the brain stem (Cools, Roberts, and Robbins 2008; Andrews et al. 2015). Similar to dopamine, most immune cells express serotonergic receptors (Hodo et al. 2020). Additionally, in response to inflammation T cells, mast cells, and macrophages can produce serotonin via increased expression of tryptophan hydroxylase enzymes which are necessary for the production of serotonin (O'Connell et al. 2006; Nakamura et al. 2008). γ-aminobutyric acid (GABA) is primarily responsible for mediating and regulating the inhibitory balance in the brain. GABA receptors can be divided primarily into two categories, ionotropic and metabotropic, based on their mode of action. It is important to note that the composition of GABA receptors decides its sensitivity to different binding agents. For instance, the $GABA_A$ receptors containing the $\gamma 2$ subunit are sensitive to benzodiazepines, whereas the $GABA_A$ receptors containing the δ subunit are particularly sensitive to neurosteroids (Rupprecht et al. 2010; Rainer et al. 2022; Pradhan et al. 2023). Therefore the composition of the GABA receptors plays a crucial role in determining the kinetics as well as the mode of action. L-glutamate is one of the primary excitatory neurotransmitters present in the CNS (Hodo et al. 2020). Glutamate is synthesized from glutamine by the enzyme glutaminase (Hodo et al. 2020). Similar to the GABA receptors, glutamatergic receptors can be divided into either metabotropic or ionotropic receptors. The metabotropic receptors act via the GPCRs and the secondary messenger system. The ionotrophic receptors on the other hand are regulated via the change in membrane potential. Both myeloid and lymphoid cells express the glutamatergic receptors and play a crucial role in mediating either pro- or anti-inflammatory response. Antigen presenting cells such as macrophages express glutamate transporters (Fallarino et al. 2010; Eisenkraft, Falk, and

Finkelstein 2013; Kritis et al. 2015; Hodo et al. 2020). Both glutamate and GABA, which are expressed in the immune cells, play a crucial role in regulating the excitatory and inhibitory balance in the brain.

1.3 CROSSTALK OF THE HPA AXIS AND THE IMMUNE SYSTEM

The immune system's role was initially suggested by Blalock (Jeffrey Edwin Blalock 1984), who proposed that it could be seen as a sixth sense, capable of detecting pathogens or tissue damage and conveying this information to the brain (J Edwin Blalock and Smith 2007). The discovery that glucocorticoids could suppress immune function unveiled the communication between the nervous and immune systems, implicating the hypothalamus-pituitary-adrenocortical (HPA) axis, influenced by the brain (Claman 1972). The HPA axis, consisting of the hypothalamus, pituitary gland, and adrenal glands, is a central player in the body's response to stress. When an individual encounters a stressful situation, whether stemming from external threats or internal stressors, the HPA axis springs into action, releasing cortisol, a pivotal stress hormone, through an intricate cascade of events that allows the body to effectively manage stress and adapt to changing circumstances. Crucially, the HPA axis does not operate in isolation; it interacts with the immune system, exemplifying the intricate interplay between neuroendocrine and immune pathways. This communication is indispensable for orchestrating the body's response to stress, modulating immune function, and upholding overall health and homeostasis. Gaining insights into the dynamic relationship between the HPA axis and the immune system offers a deeper understanding of the regulation of inflammatory responses, susceptibility to disease, and the broader repercussions of stress on health.

1.3.1 Stress and the HPA Axis

Stress triggers a range of behavioral and physical reactions aimed at adapting to the stressor. Genetic factors and coping strategies significantly influence this stress response. Stress has a profound impact on the regulation and potential dysregulation of the hypothalamic-pituitary-adrenal (HPA) axis and its interactions with other brain circuits. The HPA axis, the autonomic nervous system (ANS), and the immune system work in tandem, coordinating hormonal and inflammatory stress responses (Amsterdam, Tajima, and Sasson 2002). When an individual encounters stress, whether from external threats or internal emotional factors, the HPA axis is activated, leading to the release of cortisol and other stress-related hormones into the bloodstream. This axis serves as a vital link between the nervous and immune systems.

Both alcohol and stress stimulate nerve cells in the hypothalamus to release corticotropin-releasing hormone (CRH). CRH, within the hypothalamus, triggers the production of a morphine-like hormone, β-endorphin. CRH also travels to the anterior pituitary gland, where it prompts the production of proopiomelanocortin (POMC). POMC serves as the precursor for several stress-related hormones, including adrenocorticotropic hormone (ACTH), β-lipotropin (β-LPH), and β-endorphin. ACTH, in turn, signals the adrenal glands to produce and release cortisol. When cortisol levels reach a certain threshold, the release of CRH and ACTH diminishes. Additionally, other neurons releasing neurotransmitters like serotonin (5-HT), norepinephrine (NE), gamma-aminobutyric acid (GABA), or endogenous opioids play a role in regulating CRH release (Stephens and Wand 2012).

Overall, this process reflects the body's intricate response to stress, involving various physiological and neural systems to maintain homeostasis. Stress can both protect and potentially harm the body, depending on the duration and nature of the stressor, leading to long-term changes in the HPA axis (E Ron De Kloet, Joëls, and Holsboer 2005). This delicate balance is influenced by factors like genetic variations and the activity of proteins like FKBP5, which plays a role in regulating glucocorticoid receptor sensitivity (Binder 2009). Dysregulation in these systems may contribute to various neuropsychiatric conditions and affect overall health.

Furthermore, this stress response is not limited to a specific type of stressor; it also extends to handling seasonal changes. The duration of daily illumination, known as photoperiod, has a substantial impact on the body's rhythms, influencing the circadian patterns of cortisol and proinflammatory mediators. During the winter season, these patterns exhibit higher amplitudes and peaks later in the day, a response associated with the reduced photoperiod. Consequently, these seasonal adjustments contribute to an intensified immune response during times of acute stress (Pierre, Schlesinger, and Androulakis 2016).

Stress and the HPA axis are intricately linked, playing a crucial role in maintaining physiological balance and coping with external challenges. When an individual encounters a stressful stimulus, the body initiates a complex response aimed at restoring equilibrium and promoting survival. Cortisol, the primary glucocorticoid hormone, plays a central role in conserving energy and maintaining energy supply. This includes reducing inflammation, restricting growth, producing energy, and removing malfunctioning cellular components (Juszczak and Stankiewicz 2018). This stress response affects multiple organ systems, including the cardiovascular, respiratory, endocrine, gastrointestinal, nervous, muscular, and reproductive systems. The rapid release of cortisol under stress has several functions: it enhances gluconeogenesis, inhibits insulin production, and triggers vasoconstriction, directing blood flow to vital organs and muscles (Viblanc et al. 2018). In the brain, cortisol improves decision-making, alertness, and cognitive function, but it also has proinflammatory effects, making the body more responsive to infections (Cruz-Topete and Cidlowski 2014).

While short-term cortisol release is adaptive, chronic exposure to stress leads to detrimental effects, including metabolic disturbances, such as the development of metabolic syndrome, obesity, and insulin resistance. Chronic stress also increases susceptibility to infections, cardiovascular disease, and mental health disorders, including depression (E Ronald De Kloet et al. 1998; Miller, Cohen, and Ritchey 2002; Biddie, Conway-Campbell, and Lightman 2012; Baron and Reid 2014; Potter et al. 2016). The intricate relationship between stress, the HPA axis, and the immune system is influenced by chronic cortisol exposure. It can lead to glucocorticoid resistance, impairing the body's ability to regulate inflammatory responses effectively and creating a vicious cycle where inflammation exacerbates stress and vice versa (Miller, Cohen, and Ritchey 2002; Cohen et al. 2012).

Chronic stress and psychiatric disorders often involve HPA axis dysregulation, which existing models fail to clarify. A recent study presented a mathematical model for the HPA axis during stress, underscoring the significance of gland mass dynamics in preserving stability. Dysregulation of cortisol and ACTH was identified in conditions like alcoholism, anorexia, and postpartum, with impaired glucocorticoid receptor feedback worsening this dysregulation. These results provide valuable insights into the pathophysiology of stress-related disorders and emphasize the role of gland mass dynamics (Karin et al. 2020).

Further, another study found that chronic stress and dysregulated HPA axis function were associated with changes in DNA methylation of immune-related genes. Specifically, individuals with blunted HPA axis negative feedback showed lower DNA methylation at certain CpG sites in genes related to TNF and FKBP5. Caregivers, who experienced chronic stress, also exhibited lower methylation at CpG sites in the IL6 gene. Notably, lower TNF methylation was linked to increased TNF mRNA production after immune stimulation in peripheral blood mononuclear cells. These findings suggest that chronic stress and altered HPA axis regulation can lead to epigenetic changes in immune-related genes, potentially contributing to heightened inflammation and disease susceptibility (Palma-Gudiel et al. 2021).

1.3.2 Differential Activation Mechanism of the HPA Axis

The body's response to physical or psychological stressors involves a coordinated effort of various systems, including the sympathetic-adreno-medullar (SAM) axis and the HPA axis (Mifsud and Reul 2018). Stressors, which disrupt homeostasis, trigger these responses, resulting in physiological and behavioral changes collectively known as the stress response. The HPA axis is responsible for the slow stress response. CRH's interaction with Cortisol Releasing Hormone Binding Protein (CRH-BP) is essential for controlling the bioavailability of CRH (Kudlacek et al. 2017). During exposure to stress, CRH-BP expression increases over time, likely as a negative feedback mechanism to regulate CRH interaction with corticotropin-releasing hormone R1 receptors (CRH-R1). Notably, only unbound cortisol is biologically active (Ketchesin, Stinnett, and Seasholtz 2017).

The activation of the HPA axis is influenced by the duration and type of stress. Acute stress leads to a rapid HPA response, which typically starts with a surge in ACTH within minutes. The intensity and duration of this initial ACTH release depend on the stressor's characteristics and feedback mechanisms. However, the production and release of glucocorticoids, like corticosterone, take longer due to the adrenal's need to create these hormones. As a result, there is a time delay between the peak of ACTH and the peak of corticosterone, usually around 30 to 60 minutes. The timing of both ACTH and corticosterone responses varies based on the nature and strength of the stressor. Additionally, the duration of the corticosterone response, which can last around 2 hours, is influenced by factors such as delayed feedback mechanisms and the regulation of glucocorticoid metabolism. These temporal dynamics of the HPA axis response play a crucial role in determining the effects of stress on the body at both cellular and systemic levels (Herman 2013).

The connection between the immune system and the HPA axis became evident when infections were linked to HPA axis activation. This link was solidified when it was shown that IL-1, a cytokine produced by various immune cells in response to infections, could significantly stimulate the HPA axis (H. Besedovsky et al. 1986). IL-1 was shown to activate CRH-containing cells in the paraventricular nucleus (PVN), a crucial initial step in HPA axis activation, in many studies. In some cases, IL-1 could also directly induce ACTH secretion from the anterior pituitary and glucocorticoid release from the adrenal cortex, although these routes are not the primary mechanisms under normal physiological conditions (H. O. Besedovsky and del Rey 1996; Turnbull and Rivier 1999a; Dunn 2000).

The characteristics of different stressors play a pivotal role in shaping the magnitude and duration of the HPA axis stress responses. Stressor intensity is a critical determinant in defining the overall trajectory of the HPA axis response. Stressors with high intensity and extended duration, such as inflammatory stimuli, typically trigger prolonged responses. For instance, stimuli like a single injection of lipopolysaccharide (an endotoxin) or interleukin 1 beta lead to lasting elevations of ACTH and corticosterone, with peak levels occurring at around 1-2 hours and 1 hour, respectively (Schotanus, Tilders, and Berkenbosch 1993; Turnbull and Rivier 1999b).

In contrast, shorter responses are typically observed when exposed to psychological stressors, like unfamiliar or threatening situations. For example, in rodents, restraint stress, a common psychological stressor, leads to a peak in ACTH within approximately 15-20 minutes after the initiation of stress, with peak corticosterone responses occurring between 30-60 minutes. These responses often terminate within 2-3 hours of the stress onset (Vahl et al. 2005; Choi et al. 2007; Solomon et al. 2015). Importantly, in the case of restraint, recovery begins while the stressor is still present, indicating that the initial response starts to wane even before the animal is removed from the stressful situation. Stressors of lesser perceived severity, such as a 5-minute exposure to a novel open field, produce lower peak HPA activation (Furay, Bruestle, and Herman 2008).

It's essential to note that the mechanisms governing the activation and inhibition of the HPA axis differ among stressors. Stressors that signal threats to the body's physiological balance, such as systemic inflammation or hypoglycemia, directly communicate with CRH neurons in the PVN of the hypothalamus through sensory pathways (Herman 2013). On the other hand, psychological stimuli typically require the involvement of intermediary inhibitory nuclei, relaying information from limbic structures to the PVN. These limbic structures integrate sensory information and potential threat-related memories to generate a response in anticipation of a plausible threat (Herman et al. 2003; Radley and Sawchenko 2011).

However, the overall temporal course of the HPA axis response doesn't seem to depend on the type of stressor. For example, mice exposed to periods of restraint or hypoxia display corticosterone responses with similar peak levels and durations (Furay, Bruestle, and Herman 2008). This observation suggests that the net signal received by the PVN is the key factor determining HPA axis activation, rather than the nature of the stimulus.

The regulation of the HPA axis involves the participation of pituitary adenylate cyclase-activating polypeptide (PACAP). PACAP appears to have a multifaceted role within the HPA axis, contributing to the production of CRH and exerting a modulatory influence at various levels of this stress-responsive system. Additionally, there is evidence to suggest that PACAP is involved in the autonomic stress response, potentially leading to an increased secretion of catecholamines (King, Toufexis, and Hammack 2017). PACAP receptors function as G-protein-coupled receptors, with PACAP-R1 being the most abundant type found in both central and peripheral tissues. Moreover, PACAP might play a role in modulating the influence of estrogen on the amplification of the acute stress response (Kudlacek et al. 2017).

1.3.3 Impact of Glucocorticoid on Immune Response

The initial evidence of interactions between the HPA axis and the immune system emerged in the late 1940s when it was observed that glucocorticoids have immunosuppressive effects (Claman 1972). One notable outcome of this interaction was the shrinkage of the thymus gland, a phenomenon known to occur in stressed animals for many years. This effect is due to the apoptosis of immature lymphocytes in the thymus. Additionally, glucocorticoids can prompt some lymphocytes to exit the thymus and enter the general circulation, potentially for immune defense purposes. Notably, glucocorticoids can induce lymphocyte death at concentrations only slightly higher than those found in stressed animals. The interaction primarily starts with glucocorticoid binding to the glucocorticoid receptor (GR). This binding indirectly influences the activity of specific intracellular proteins critical for immune cell function (Auphan et al. 1995; De Bosscher, Berghe, and Haegeman 2000). In addition, glucocorticoids are known for their lympholytic properties and are used in treating hematological malignancies, inducing apoptosis in lymphocytes (Herold, McPherson, and Reichardt 2006; Boldizsar et al. 2010). Notably, various immune cells beyond lymphocytes are sensitive to glucocorticoid-induced apoptosis, adding to the diversity of glucocorticoid effects on immune function (Weinstein et al. 1998; Kim et al. 2001; Druilhe, Letuve, and Pretolani 2003).

Glucocorticoids exert profound effects on cellular immunity, particularly on T-cell responses. A study by Talaber G. et al., questions the conventional belief that glucocorticoids directly influence thymocyte regulation and proposes that ACTH, instead, plays a role in governing thymic output by acting on thymic epithelial cells (Talaber, Tuckermann, and Okret 2015). Further, glucocorticoids influence various DC types, diminishing their activity by inhibiting maturation and downregulating the expression of key molecules involved in T cell activation, such as MHC class II, co-stimulatory molecules, and pro-inflammatory cytokines (Szatmari and Nagy 2008). T cell activation is also hampered by glucocorticoids, affecting TCR signaling and reducing proliferative responses and cytokine production, notably IL-2 (Gillis, Crabtree, and Smith 1979; Tsitoura and Rothman 2004; Löwenberg et al. 2007). Glucocorticoids exhibit a nuanced impact on different T helper (TH) cell subsets. They preferentially suppress TH1 and TH17 cell responses, potentially sparing or even promoting TH2 and regulatory T (Treg) cells. The mechanisms underlying these preferences are multifaceted, including effects on cytokine production, receptor expression, and transcription factors (Elenkov 2004; Liberman et al. 2007). This complex regulation of T-cell responses contributes to the multifaceted role of glucocorticoids in immune and inflammatory disorders, making them valuable in the treatment of various conditions.

The influence of glucocorticoids on B lymphopoiesis is reminiscent of their impact on thymopoiesis. Immature B cells are more susceptible to glucocorticoid-induced apoptosis, compared to mature B cells (Garvy et al. 1993; Igarashi et al. 2005). Insights into the direct effects of glucocorticoids on B cell function can be gleaned from studies involving the glucocorticoid target gene Glucocorticoid-Induced Leucine Zipper (Gilz). B cells lacking Gilz exhibit heightened proliferative responses when stimulated through the BCR, and these mice can develop a lupus-like syndrome with age (Jones et al. 2016). A B cell-specific deficiency of Gilz leads to increased B cell numbers systemically, suggesting an intrinsic role for glucocorticoid receptor signaling in B cell survival and homeostasis (Bruscoli et al. 2015). Glucocorticoids can increase IgE in asthma patients, but their clinical benefits in asthma and allergies indicate no exacerbation of IgE production (Posey et al. 1978; Zieg et al. 1994). Additionally, endogenous glucocorticoids may enhance antibody responses to immunization (Fleshner et al. 2001).

Glucocorticoids serve as potent regulators of inflammation, influencing the three main phases of the inflammatory response. In the initial alarm phase, glucocorticoids mitigate the activation of pattern recognition receptors (PRRs), Fcε receptors (FcεRs), and cytokine receptors, dampening the production of key inflammatory mediators. During the subsequent mobilization phase, they reduce the expression of adhesion molecules, chemokines, and integrins on vascular endothelium, limiting the recruitment of immune cells into the inflamed tissue. In the resolution phase, glucocorticoids promote the differentiation of "M2c" macrophages, which aid in clearing apoptotic immune cells and release anti-inflammatory factors (Cain and Cidlowski 2017).

While glucocorticoids are widely recognized for their anti-inflammatory and immunosuppressive properties, recent research has unveiled their ability to enhance inflammation and immunity under specific conditions. The direction of these effects appears to be dose-dependent. In certain studies, high doses of corticosterone inhibited the transcription of inflammatory genes in macrophages activated by lipopolysaccharide (LPS) and IFNγ, whereas low doses of corticosterone enhanced inflammatory gene expression (Lim et al. 2007). Additionally, the timing of glucocorticoid exposure in relation to a noxious stimulus can influence their impact on inflammation. In rats, glucocorticoids administered before the LPS challenge enhanced pro-inflammatory cytokine expression, while post-LPS treatment dampened it (Frank et al. 2010). Similarly, acute low-dose corticosterone treatment before challenge increased inflammation in a rat model of delayed-type hypersensitivity, whereas high-dose or prolonged glucocorticoid treatment mitigated the response (Dhabhar and McEwen 1999).

Furthermore, genome-wide expression studies have revealed that glucocorticoids tend to spare or even enhance gene pathways associated with innate immunity, while selectively suppressing pathways linked to adaptive immunity. For example, glucocorticoids boost the expression of pattern recognition receptors (PRRs), scavenger receptors, cytokine and chemokine receptors, and complement components involved in innate immunity, yet they suppress genes encoding T cell receptor (TCR) components, MHC class II, and co-stimulatory molecules associated with T cell activation (Galon et al. 2002).

1.4 IMMUNE CELL TRAFFICKING AND THE IMMUNE PRIVILEGE OF THE CNS

Imagine the huge risk to the CNS arising from the foreign particles and pathogens, risk factors circulating in the blood, and the immune cells. However, CNS is safely guarded against all these factors by the very well-known "blood-brain barrier" (BBB) which is tightly regulated and mediated by the astrocytic endfeet processes, tight junctions between endothelial cells and the basal lamina (Ransohoff, Kivisäkk, and Kidd 2003; Bentivoglio and Kristensson 2014; A Louveau, Harris, and Kipnis 2015a). The reference of CNS immune privilege was defined early, based on the isolated environment from the periphery and the slow immune reactions in the brain (A Louveau, Harris, and Kipnis 2015a). As time progressed, the immune interaction of the brain remained unique but the definition of immune privilege changed over time. Limited classical lymphatic drainage of CNS antigens is partly thought to contribute to the immune privilege system of the brain (A Louveau, Harris, and Kipnis 2015a).

The CNS consists of interstitial fluid (ISF) and cerebrospinal fluid (CSF), in addition to blood (A Louveau, Harris, and Kipnis 2015a). The choroid plexus produces CSF, which circulates from the lateral ventricles to the third and fourth ventricles, finally reaching the subarachnoid space within the meninges in the subarachnoid space. Circulation of ISF occurs within the cerebral parenchyma (Weller et al. 2009; Laman and Weller 2013). Previous studies have highlighted that CSF drains into deep cervical lymph via the cribriform plate into the lymphatic system of the nasal mucosa (Harling-Berg et al. 1989; Cserr, Harling-Berg, and Knopf 1992). Additionally, recent studies highlight the role of the meningeal lymphatic system, a comprehensive functional lymphatic system located in the meninges that facilitates the drainage of immune cells, fluids, and macromolecules to deep cervical nodes from the CNS (Aspelund et al. 2015; Antoine Louveau et al. 2015b). This was primarily illustrated via the injection of tracers and macromolecules in the brain which were shown to drain into the deep cervical nodes (Marshall L Rennels et al. 1985; M L Rennels, Blaumanis, and Grady 1990). There appears to be a physical connection between the meningeal lymphatic vessels and the deep cervical lymph nodes, as damage of the lymphatic connections to the deep cervical lymph nodes induced a lymphedema-like sensation on the meningeal lymphatics (Antoine Louveau et al. 2015b). Under normal conditions, T cells, B cells, and dendritic cells were present in the meningeal lymphatic vessels, indicating their involvement in immune cell trafficking outside the CNS (Antoine Louveau et al. 2015b). It is important also to note that the protein aggregates can be drained to the periphery from the CNS via the meningeal lymphatic vessels (A Louveau, Harris, and Kipnis 2015a). Additionally, reports suggest an important physiological role of immune cells in the cognitive process. For instance, T cells are usually associated with cognitive processes in a physiological state

(A Louveau, Harris, and Kipnis 2015a). T-cell deficiency was previously associated with impaired spatial memory and learning (Brynskikh et al. 2008; Derecki et al. 2010; Radjavi, Smirnov, and Kipnis 2014). Similarly, impaired cognitive performance is observed when there is an acute depletion of T cells or impaired migration of CNS tissue (Wolf et al. 2009; Derecki et al. 2010). Taken together, this suggests a unique immune response in the CNS different from the periphery.

1.5 CLINICAL RELEVANCE AND THERAPEUTICS POTENTIAL OF NEUROIMMUNE NETWORK

The cross-talk between neurons and immune cells is being clinically explored to develop efficient therapeutics. Chronic autoimmune disorders such as rheumatoid arthritis affect the global population and damage the economy and resources. Multiple sclerosis (MS), is a pathological scenario, where the immune cells affect the CNS, resulting in paralysis (Voet, Prinz, and van Loo 2019). When autoimmune T cells travel to the brain and cross the BBB, they cause damage to the brain. Therapeutics targeting the T cell migration to the CNS are being developed in experimental autoimmune encephalomyelitis (EAE) animal models of MS (Constantinescu et al. 2011). Similarly, dementia-related disorders are widely associated with hyperactive immune systems.

In progressive neurodisorders such as Alzheimer's (AD), reactive gliosis is commonly associated with deteriorating pathophysiology. Loss of functional synapses is effectively associated with cognitive decline in AD. Aberrant phagocytic action of glial cells, including the astrocytes and microglia, plays a crucial role in the reduction of dendritic spines, a common pathophysiological response commonly associated with AD. Targeting the different pathophysiological responses of neurodisorders is being utilized in the development of therapeutics. For instance, in AD, the microglia shift from anti to pro-inflammatory cytokines (Q. Wang et al. 2021; Guo, Wang, and Yin 2022). Given, the neuroprotective role of anti-inflammatory cytokine therapeutics targeting this shift are being tested in different animal models (Asanuma, Miyazaki, and Ogawa 2004; Skaper 2007). Similarly, the Aβ plaques are often thought to sequester toxic Aβ oligomeric species, and astrocytic processes are highly associated with Aβ plaques (Koffie et al. 2009; Karisetty et al. 2020). Recent reports suggest that the astrocytes that are close to and are associated with the amyloid plaques have reduced astrocytic phagocytosing capacity and follow an A1 astrocytic fate (Sakakibara et al. 2019). This in general, differs from the astrocytes, which are actively phagocytosing and engulfing the synapses. Thus, astrocytes display differential phagocytic action based on the region and distance from the plaques. A hyperactive complement system commonly associated with neurodisorders is also being explored as a possible target for the development of therapeutics. The complement system acts as a surveillance system and provides the first line of defense against foreign bodies by executing an innate immune response. In several disorders, such as Parkinson's and AD, a hyperactive complement system is often thought to drive the toxic pathological effect (discussed in detail in the Chapter "**Neuroimmune System Scenario in Neurodegenerative Disorders**").

REFERENCES

Amsterdam, Abraham, Kimihisa Tajima, and Ravid Sasson. 2002. "Cell-specific regulation of apoptosis by glucocorticoids: implication to their anti-inflammatory action." *Biochemical Pharmacology* 64 (5–6): 843–50.

Andrews, Paul W, Aadil Bharwani, Kyuwon R Lee, Molly Fox, and J Anderson Thomson Jr. 2015. "Is serotonin an upper or a Downer? The evolution of the serotonergic system and its role in depression and the antidepressant response." *Neuroscience & Biobehavioral Reviews* 51: 164–88.

Ansel, John C, Jeffrey R Brown, Donald G Payan, and Melissa A Brown. 1993. "Substance P selectively activates TNF-alpha gene expression in murine mast cells." *Journal of Immunology (Baltimore, Md.: 1950)* 150 (10): 4478–85.

Asanuma, M, I Miyazaki, and N Ogawa. 2004. "Neuroprotective effects of nonsteroidal anti-inflammatory drugs on neurodegenerative diseases." *Current Pharmaceutical Design* 10 (6): 695–700.

Aspelund, Aleksanteri, Salli Antila, Steven T Proulx, Tine Veronica Karlsen, Sinem Karaman, Michael Detmar, Helge Wiig, and Kari Alitalo. 2015. "A dural lymphatic vascular system that drains brain interstitial fluid and macromolecules." *Journal of Experimental Medicine* 212 (7): 991–99.

Auphan, Nathalie, Joseph A DiDonato, Caridad Rosette, Arno Helmberg, and Michael Karin. 1995. "Immunosuppression by glucocorticoids: Inhibition of NF-KB activity through induction of IκB synthesis." *Science* 270 (5234): 286–90.

Banisadr, Ghazal, William Rostène, Patrick Kitabgi, and Stephane M Parsadaniantz. 2005. "Chemokines and brain functions." *Current Drug Targets-Inflammation & Allergy* 4 (3): 387–99.

Baron, Kelly Glazer, and Kathryn J Reid. 2014. "Circadian misalignment and health." *International Review of Psychiatry* 26 (2): 139–54.

Barrios, Juliana, Kruti R Patel, Linh Aven, Rebecca Achey, Martin S Minns, Yoonjoo Lee, Vickery E Trinkaus-Randall, and Xingbin Ai. 2017. "Early life allergen-induced mucus overproduction requires augmented neural stimulation of pulmonary neuroendocrine cell secretion." *The FASEB Journal* 31 (9): 4117.

Becher, Burkhard, Sabine Spath, and Joan Goverman. 2017. "Cytokine networks in neuroinflammation." *Nature Reviews Immunology* 17 (1): 49–59.

Bentivoglio, Marina, and Krister Kristensson. 2014. "Tryps and trips: Cell trafficking across the 100-year-old blood–brain barrier." *Trends in Neurosciences* 37 (6): 325–33.

Berthoud, Hans-Rudolf, and Winfried L Neuhuber. 2000. "Functional and chemical anatomy of the afferent vagal system." *Autonomic Neuroscience* 85 (1–3): 1–17.

Besedovsky, Hugo Oscar, and Adriana del Rey. 1996. "Immune-neuro-endocrine interactions: Facts and hypotheses." *Endocrine Reviews* 17 (1): 64–102.

Besedovsky, Hugo, Adriana Del Rey, Ernst Sorkin, and Charles A Dinarello. 1986. "Immunoregulatory feedback between interleukin-1 and glucocorticoid hormones." *Science* 233 (4764): 652–54.

Besser, Michal J, Yonatan Ganor, and Mia Levite. 2005. "Dopamine by itself activates either D2, D3 or D1/D5 dopaminergic receptors in normal human T-Cells and triggers the selective secretion of either IL-10, TNFα or Both." *Journal of Neuroimmunology* 169 (1–2): 161–71.

Biddie, Simon C, Becky L Conway-Campbell, and Stafford L Lightman. 2012. "Dynamic regulation of glucocorticoid signalling in health and disease." *Rheumatology* 51 (3): 403–12.

Binder, Elisabeth B. 2009. "The role of FKBP5, a Co-Chaperone of the glucocorticoid receptor in the pathogenesis and therapy of affective and anxiety disorders." *Psychoneuroendocrinology* 34: S186–95.

Blalock, J Edwin, and Eric M Smith. 2007. "Conceptual development of the immune system as a sixth sense." *Brain, Behavior, and Immunity* 21 (1): 23–33.

Blalock, Jeffrey Edwin. 1984. "The immune system as a sensory organ." *Journal of Immunology (Baltimore, Md.: 1950)* 132 (3): 1067–70.

Boldizsar, Ferenc, Gergely Talaber, Mariann Szabo, Domokos Bartis, Laszlo Palinkas, Peter Nemeth, and Timea Berki. 2010. "Emerging pathways of non-genomic glucocorticoid (GC) signalling in T cells." *Immunobiology* 215 (7): 521–26.

Bosscher, Karolien De, Wim Vanden Berghe, and Guy Haegeman. 2000. "Mechanisms of anti-inflammatory action and of immunosuppression by glucocorticoids: Negative interference of activated glucocorticoid receptor with transcription factors." *Journal of Neuroimmunology* 109 (1): 16–22.

Boulanger, Lisa M. 2009. "Immune proteins in brain development and synaptic plasticity." *Neuron* 64 (1): 93–109.

Bruscoli, Stefano, Michele Biagioli, Daniele Sorcini, Tiziana Frammartino, Monica Cimino, Paolo Sportoletti, Emanuela Mazzon, Oxana Bereshchenko, and Carlo Riccardi. 2015. "Lack of Glucocorticoid-Induced Leucine Zipper (GILZ) Deregulates B-cell survival and results in B-cell lymphocytosis in mice." *Blood, The Journal of the American Society of Hematology* 126 (15): 1790–1801.

Brynskikh, Anna, Ted Warren, Julia Zhu, and Jonathan Kipnis. 2008. "Adaptive immunity affects learning behavior in mice." *Brain, Behavior, and Immunity* 22 (6): 861–69.

Buffington, Shelly A, Gonzalo Viana Di Prisco, Thomas A Auchtung, Nadim J Ajami, Joseph F Petrosino, and Mauro Costa-Mattioli. 2016. "Microbial reconstitution reverses maternal diet-induced social and synaptic deficits in offspring." *Cell* 165 (7): 1762–75.

Cain, Derek W, and John A Cidlowski. 2017. "Immune regulation by glucocorticoids." *Nature Reviews Immunology* 17 (4): 233–47.

Chan, Kenny L, Wolfram C Poller, Filip K Swirski, and Scott J Russo. 2023. "Central regulation of stress-evoked peripheral immune responses." *Nature Reviews Neuroscience*, 1–14.

Choi, Dennis C, Amy R Furay, Nathan K Evanson, Michelle M Ostrander, Yvonne M Ulrich-Lai, and James P Herman. 2007. "Bed nucleus of the stria terminalis subregions differentially regulate hypothalamic–pituitary–adrenal axis activity: Implications for the integration of limbic inputs." *Journal of Neuroscience* 27 (8): 2025–34.

Claman, Henry N. 1972. "Corticosteroids and Lymphoid cells." *New England Journal of Medicine* 287 (8): 388–97.

Cohen, Sheldon, Denise Janicki-Deverts, William J Doyle, Gregory E Miller, Ellen Frank, Bruce S Rabin, and Ronald B Turner. 2012. "Chronic stress, glucocorticoid receptor resistance, inflammation, and disease risk." *Proceedings of the National Academy of Sciences* 109 (16): 5995–99.

Constantinescu, Cris S, Nasr Farooqi, Kate O'Brien, and Bruno Gran. 2011. "Experimental Autoimmune Encephalomyelitis (EAE) as a Model for Multiple Sclerosis (MS)." *British Journal of Pharmacology* 164 (4): 1079–1106.

Cools, Roshan, Angela C Roberts, and Trevor W Robbins. 2008. "Serotoninergic regulation of emotional and behavioural control processes." *Trends in Cognitive Sciences* 12 (1): 31–40.

Cosentino, Marco, Raffaella Bombelli, Marco Ferrari, Franca Marino, Emanuela Rasini, Georges J M Maestroni, Ario Conti, Monica Boveri, Sergio Lecchini, and Gianmario Frigo. 2000. "HPLC-ED measurement of endogenous catecholamines in human immune cells and hematopoietic cell lines." *Life Sciences* 68 (3): 283–95.

Cosentino, Marco, Anna Maria Fietta, Marco Ferrari, Emanuela Rasini, Raffaella Bombelli, Elena Carcano, Federica Saporiti, Federica Meloni, Franca Marino, and Sergio Lecchini. 2007. "Human CD4+ CD25+ Regulatory T cells selectively express tyrosine hydroxylase and contain endogenous catecholamines subserving an autocrine/paracrine inhibitory functional loop." *Blood* 109 (2): 632–42.

Costello, Richard W, Brian H Schofield, Gail M Kephart, Gerald J Gleich, David B Jacoby, and Allison D Fryer. 1997. "Localization of eosinophils to airway nerves and effect on neuronal m2 muscarinic receptor function." *American Journal of Physiology-Lung Cellular and Molecular Physiology* 273 (1): L93–103.

Cruz-Topete, Diana, and John A Cidlowski. 2014. "One hormone, two actions: Anti-and pro-inflammatory effects of glucocorticoids." *Neuroimmunomodulation* 22 (1–2): 20–32.

Cserr, Helen F, Christine J Harling-Berg, and Paul M Knopf. 1992. "Drainage of brain extracellular fluid into blood and deep cervical lymph and its immunological significance." *Brain Pathology* 2 (4): 269–76.

Cyphert, Jaime M, Martina Kovarova, Irving C Allen, John M Hartney, Dennis L Murphy, Jurgen Wess, and Beverly H Koller. 2009. "Cooperation between mast cells and neurons is essential for antigen-mediated bronchoconstriction." *The Journal of Immunology* 182 (12): 7430–39.

Derecki, Noël C, Amber N Cardani, Chun Hui Yang, Kayla M Quinnies, Anastasia Crihfield, Kevin R Lynch, and Jonathan Kipnis. 2010. "Regulation of learning and memory by meningeal immunity: A key role for IL-4." *Journal of Experimental Medicine* 207 (5): 1067–80.

Dhabhar, Firdaus S, and Bruce S McEwen. 1999. "Enhancing versus suppressive effects of stress hormones on skin immune function." *Proceedings of the National Academy of Sciences* 96 (3): 1059–64.

Domingues, Helena S, Camila C Portugal, Renato Socodato, and João B Relvas. 2016. "Oligodendrocyte, astrocyte, and microglia crosstalk in myelin development, damage, and repair." *Frontiers in Cell and Developmental Biology* 4: 71.

Druilhe, A, S Letuve, and M Pretolani. 2003. "Glucocorticoid-induced apoptosis in human Eosinophils: Mechanisms of action." *Apoptosis* 8: 481–95.

Dunn, Adrian J. 2000. "Cytokine activation of the HPA axis." *Annals of the New York Academy of Sciences* 917 (1): 608–17.

Eisenkraft, Arik, Avshalom Falk, and Arseny Finkelstein. 2013. "The role of glutamate and the immune system in organophosphate-induced CNS damage." *Neurotoxicity Research* 24: 265–79.

Elenkov, Ilia J. 2004. "Glucocorticoids and the Th1/Th2 balance." *Annals of the New York Academy of Sciences* 1024 (1): 138–46.

Fallarino, Francesca, Claudia Volpi, Francesco Fazio, Serena Notartomaso, Carmine Vacca, Carla Busceti, Silvio Bicciato, Giuseppe Battaglia, Valeria Bruno, and Paolo Puccetti. 2010. "Metabotropic glutamate receptor-4 modulates adaptive immunity and restrains neuroinflammation." *Nature Medicine* 16 (8): 897–902.

Fleshner, Monika, Terrence Deak, Kien T Nguyen, Linda R Watkins, and Steven F Maier. 2001. "Endogenous glucocorticoids play a positive regulatory role in the anti-keyhole limpet hemocyanin in vivo antibody response." *The Journal of Immunology* 166 (6): 3813–19.

Frank, Matthew G, Zurine D Miguel, Linda R Watkins, and Steven F Maier. 2010. "Prior exposure to glucocorticoids sensitizes the neuroinflammatory and peripheral inflammatory responses to E. Coli Lipopolysaccharide." *Brain, Behavior, and Immunity* 24 (1): 19–30.

Furay, Amy R, Amy E Bruestle, and James P Herman. 2008. "The role of the forebrain glucocorticoid receptor in acute and chronic stress." *Endocrinology* 149 (11): 5482–90.

Galon, Jérôme, Denis Franchimont, Naoki Hiroi, Gregory Frey, Antje Boettner, Monika Ehrhart-Bornstein, John J O'Shea, George P Chrousos, and Stefan R Bornstein. 2002. "Gene profiling reveals unknown enhancing and suppressive actions of glucocorticoids on immune cells." *The FASEB Journal* 16 (1): 61–71.

Garvy, B A, L E King, W G Telford, L A Morford, and P J Fraker. 1993. "Chronic elevation of plasma corticosterone causes reductions in the number of cycling cells of the B lineage in murine bone marrow and induces apoptosis." *Immunology* 80 (4): 587.

Gesù, Claudia M Di, Lisa M Matz, Ian J Bolding, Robert Fultz, Kristi L Hoffman, Antonella Marino Gammazza, Joseph F Petrosino, and Shelly A Buffington. 2022. "Maternal gut microbiota mediate intergenerational effects of high-fat diet on descendant social behavior." *Cell Reports* 41 (2).

Gillis, Steven, Gerald R Crabtree, and Kendall A Smith. 1979. "Glucocorticoid-induced inhibition of T cell growth factor production: I. The effect on mitogen-induced Lymphocyte proliferation." *The Journal of Immunology* 123 (4): 1624–31.

Godinho-Silva, Cristina, Filipa Cardoso, and Henrique Veiga-Fernandes. 2019. "Neuro–immune cell units: A new paradigm in physiology." *Annual Review of Immunology* 37: 19–46.

Greter, Melanie, Iva Lelios, Pawel Pelczar, Guillaume Hoeffel, Jeremy Price, Marylene Leboeuf, Thomas M Kündig, Karl Frei, Florent Ginhoux, and Miriam Merad. 2012. "Stroma-derived Interleukin-34 controls the development and maintenance of langerhans cells and the maintenance of microglia." *Immunity* 37 (6): 1050–60.

Guo, Shenrui, Hui Wang, and Yafu Yin. 2022. "Microglia polarization from M1 to M2 in neurodegenerative diseases." *Frontiers in Aging Neuroscience* 14 :815347.

Gupta, Kalpna, and Ilkka T Harvima. 2018. "Mast cell-neural interactions contribute to pain and itch." *Immunological Reviews* 282 (1): 168–87.

Harling-Berg, Christine, Paul M Knopf, Jennifer Merriam, and Helen F Cserr. 1989. "Role of cervical lymph nodes in the systemic humoral immune response to human serum albumin microinfused into rat cerebrospinal fluid." *Journal of Neuroimmunology* 25 (2–3): 185–93.

Hatta, Toshihisa, Kenji Moriyama, Kinichi Nakashima, Tetsuya Taga, and Hiroki Otani. 2002. "The role of Gp130 in cerebral cortical development: In vivo functional analysis in a mouse exo uterosystem." *Journal of Neuroscience* 22 (13): 5516–24.

Herman, James P. 2013. "Neural control of chronic stress adaptation." *Frontiers in Behavioral Neuroscience* 7: 61.

Herman, James P, Helmer Figueiredo, Nancy K Mueller, Yvonne Ulrich-Lai, Michelle M Ostrander, Dennis C Choi, and William E Cullinan. 2003. "Central mechanisms of stress integration: Hierarchical circuitry controlling hypothalamo–pituitary–adrenocortical responsiveness." *Frontiers in Neuroendocrinology* 24 (3): 151–80.

Herold, M J, K G McPherson, and H M Reichardt. 2006. "Glucocorticoids in T cell apoptosis and function." *Cellular and Molecular Life Sciences* 63: 60–72.

Hodo, Thomas Wesley, Maria Teresa Prudente De Aquino, Akiko Shimamoto, and Anil Shanker. 2020. "Critical neurotransmitters in the neuroimmune network." *Frontiers in Immunology* 11: 1869.

Huh, Jun R, and Henrique Veiga-Fernandes. 2020. "Neuroimmune circuits in inter-organ communication." *Nature Reviews Immunology* 20 (4): 217–28.

Igarashi, Hideya, Kay L Medina, Takafumi Yokota, Maria Isabel D Rossi, Nobuo Sakaguchi, Philip C Comp, and Paul W Kincade. 2005. "Early lymphoid progenitors in mouse and man are highly sensitive to glucocorticoids." *International Immunology* 17 (5): 501–11.

Jacoby, David B, Richard M Costello, and Allison D Fryer. 2001. "Eosinophil recruitment to the airway nerves." *Journal of Allergy and Clinical Immunology* 107 (2): 211–18.

Jones, Sarah A, Andrew E J Toh, Dragana Odobasic, Marie-Anne Virginie Oudin, Qiang Cheng, Jacinta P W Lee, Stefan J White, Brendan E Russ, Simona Infantino, and Amanda Light. 2016. "Glucocorticoid-Induced Leucine Zipper (GILZ) Inhibits B cell activation in systemic lupus erythematosus." *Annals of the Rheumatic Diseases* 75 (4): 739–47.

Jung, Yeon-Joo, and Won-Suk Chung. 2018. "Phagocytic roles of glial cells in healthy and diseased brains." *Biomolecules & Therapeutics* 26 (4): 350.

Juszczak, Grzegorz R, and Adrian M Stankiewicz. 2018. "Glucocorticoids, genes and brain function." *Progress in Neuro-Psychopharmacology & Biological Psychiatry* 82 (March): 136–68. https://doi.org/10.1016/j.pnpbp.2017.11.020

Karin, Omer, Moriya Raz, Avichai Tendler, Alon Bar, Yael Korem Kohanim, Tomer Milo, and Uri Alon. 2020. "A new model for the HPA axis explains dysregulation of stress hormones on the timescale of weeks." *Molecular Systems Biology* 16 (7): e9510.

Karisetty, Bhanu Chandra, Akanksha Bhatnagar, Ellen M Armour, Mariah Beaver, Haolin Zhang, and Felice Elefant. 2020. "Amyloid-β peptide impact on synaptic function and neuroepigenetic gene control reveal new therapeutic strategies for Alzheimer's Disease." *Frontiers in Molecular Neuroscience* 13: 577622.

Kashem, Sakeen W, Maureen S Riedl, Chen Yao, Christopher N Honda, Lucy Vulchanova, and Daniel H Kaplan. 2015. "Nociceptive sensory fibers drive Interleukin-23 production from CD301b+ Dermal dendritic cells and drive protective cutaneous immunity." *Immunity* 43 (3): 515–26.

Ketchesin, Kyle D, Gwen S Stinnett, and Audrey F Seasholtz. 2017. "Corticotropin-releasing hormone-binding protein and stress: From invertebrates to humans." *Stress* 20 (5): 449–64.

Kim, Kwang Dong, Yong-Kyung Choe, In Seong Choe, and Jong-Seok Lim. 2001. "Inhibition of glucocorticoid-mediated, caspase-independent dendritic cell death by CD40 activation." *Journal of Leukocyte Biology* 69 (3): 426–34.

King, S Bradley, Donna J Toufexis, and Sayamwong E Hammack. 2017. "Pituitary Adenylate Cyclase Activating Polypeptide (PACAP), stress, and sex hormones." *Stress* 20 (5): 465–75.

Kloet, E Ron De, Marian Joëls, and Florian Holsboer. 2005. "Stress and the brain: From adaptation to disease." *Nature Reviews Neuroscience* 6 (6): 463–75.

Kloet, E Ronald De, Erno Vreugdenhil, Melly S Oitzl, and Marian Joëls. 1998. "Brain corticosteroid receptor balance in health and disease." *Endocrine Reviews* 19 (3): 269–301.

Koffie, Robert M, Melanie Meyer-Luehmann, Tadafumi Hashimoto, Kenneth W Adams, Matthew L Mielke, Monica Garcia-Alloza, Kristina D Micheva, Stephen J Smith, M Leo Kim, and Virginia M Lee. 2009. "Oligomeric Amyloid β associates with postsynaptic densities and correlates with excitatory synapse loss near senile plaques." *Proceedings of the National Academy of Sciences* 106 (10): 4012–17.

Koren, Omry, Julia K Goodrich, Tyler C Cullender, Aymé Spor, Kirsi Laitinen, Helene Kling Bäckhed, Antonio Gonzalez, Jeffrey J Werner, Largus T Angenent, and Rob Knight. 2012. "Host remodeling of the gut microbiome and metabolic changes during pregnancy." *Cell* 150 (3): 470–80.

Kritis, Aristeidis A, Eleni G Stamoula, Krystallenia A Paniskaki, and Theofanis D Vavilis. 2015. "Researching glutamate–induced cytotoxicity in different cell lines: A comparative/collective analysis/study." *Frontiers in Cellular Neuroscience* 9: 91.

Kudlacek, Oliver, Tina Hofmaier, Anton Luf, Felix P Mayer, Thomas Stockner, Constanze Nagy, Marion Holy, Michael Freissmuth, Rainer Schmid, and Harald H Sitte. 2017. "Cocaine adulteration." *Journal of Chemical Neuroanatomy* 83: 75–81.

Kulka, Marianna, Cecilia H Sheen, Brian P Tancowny, Leslie C Grammer, and Robert P Schleimer. 2008. "Neuropeptides activate human mast cell degranulation and chemokine production." *Immunology* 123 (3): 398–410.

Laman, Jon D, and Roy O Weller. 2013. "Drainage of CELLS AND SOLUBLE ANTIGEN from the CNS to regional lymph nodes." *Journal of Neuroimmune Pharmacology* 8: 840–56.

Leclercq, Sophie, Firoz M Mian, Andrew M Stanisz, Laure B Bindels, Emmanuel Cambier, Hila Ben-Amram, Omry Koren, Paul Forsythe, and John Bienenstock. 2017. "Low-dose penicillin in early life induces long-term changes in murine gut microbiota, brain cytokines and behavior." *Nature Communications* 8 (1): 15062.

Liberman, Ana Clara, Damian Refojo, Jimena Druker, Marta Alicia Toscano, Theo Rein, Florian Holsboer, and Eduardo Simon Arzt. 2007. "The activated glucocorticoid receptor inhibits the transcription factor T-Bet by direct protein-protein interaction." In. *Federation of American Societies for Experimental Biology*.

Lim, Hee-Young, Nora Müller, Marco J Herold, Jens Van Den Brandt, and Holger M Reichardt. 2007. "Glucocorticoids exert opposing effects on macrophage function dependent on their concentration." *Immunology* 122 (1): 47–53.

Louveau, A, T H Harris, and J Kipnis. 2015a. "Revisiting the concept of CNS immune privilege antoine." *Trends Immunol* 36 (10): 569–77.

Louveau, Antoine, Igor Smirnov, Timothy J Keyes, Jacob D Eccles, Sherin J Rouhani, J David Peske, Noel C Derecki, David Castle, James W Mandell, and Kevin S Lee. 2015b. "Structural and functional features of central nervous system lymphatic vessels." *Nature* 523 (7560): 337–41.

Löwenberg, Mark, Auke P Verhaar, Gijs R van den Brink, and Daniel W Hommes. 2007. "Glucocorticoid signaling: A nongenomic mechanism for T-cell immunosuppression." *Trends in Molecular Medicine* 13 (4): 158–63.

Malkova, Natalia V, Z Yu Collin, Elaine Y Hsiao, Marlyn J Moore, and Paul H Patterson. 2012. "Maternal immune activation yields offspring displaying mouse versions of the three core symptoms of autism." *Brain, Behavior, and Immunity* 26 (4): 607–16.

Matejuk, Agata, and Richard M Ransohoff. 2020. "Crosstalk between Astrocytes and Microglia: An overview." *Frontiers in Immunology* 11: 1416.

Metz, Christine N, and Valentin A Pavlov. 2018. "Vagus nerve cholinergic circuitry to the liver and the gastrointestinal tract in the neuroimmune communicatome." *American Journal of Physiology-Gastrointestinal and Liver Physiology* 315 (5): G651–58.

Mifsud, Karen R, and Johannes M H M Reul. 2018. "Mineralocorticoid and glucocorticoid receptor-mediated control of genomic responses to stress in the brain." *Stress* 21 (5): 389–402.

Miller, Gregory E, Sheldon Cohen, and A Kim Ritchey. 2002. "Chronic psychological stress and the regulation of pro-inflammatory cytokines: A glucocorticoid-resistance model." *Health Psychology* 21 (6): 531.

Mizuno, Tetsuya, Yukiko Doi, Hiroyuki Mizoguchi, Shijie Jin, Mariko Noda, Yoshifumi Sonobe, Hideyuki Takeuchi, and Akio Suzumura. 2011. "Interleukin-34 selectively enhances the neuroprotective effects of microglia to attenuate oligomeric Amyloid-β Neurotoxicity." *The American Journal of Pathology* 179 (4): 2016–27.

Nakamura, Kazuhiro, Taku Sato, Akiko Ohashi, Hiromichi Tsurui, and Hiroyuki Hasegawa. 2008. "Role of a serotonin precursor in development of gut microvilli." *The American Journal of Pathology* 172 (2): 333–44.

Nakano, Kazuhisa, Takehiro Higashi, Rie Takagi, Kumiko Hashimoto, Yoshiya Tanaka, and Sho Matsushita. 2009. "Dopamine released by dendritic cells polarizes Th2 differentiation." *International Immunology* 21 (6): 645–54.

Nilsson, G, K Alving, S Ahlstedt, T Hökfelt, and J M Lundberg. 1990. "Peptidergic innervation of rat lymphoid tissue and lung: relation to mast cells and sensitivity to capsaicin and immunization." *Cell and Tissue Research* 262: 125–33.

O'Connell, Peta J, Xiangbin Wang, Matilde Leon-Ponte, Corrie Griffiths, Sandeep C Pingle, and Gerard P Ahern. 2006. "A novel form of immune signaling revealed by transmission of the inflammatory mediator serotonin between dendritic cells and T cells." *Blood* 107 (3): 1010–17.

Oetjen, Landon K, Madison R Mack, Jing Feng, Timothy M Whelan, Haixia Niu, Changxiong J Guo, Sisi Chen, Anna M Trier, Amy Z Xu, and Shivani V Tripathi. 2017. "Sensory neurons co-opt classical immune signaling pathways to mediate chronic itch." *Cell* 171 (1): 217–28.

Ohara, Peter T, Jean-Philippe Vit, Aditi Bhargava, Marcela Romero, Christopher Sundberg, Andrew C Charles, and Luc Jasmin. 2009. "Gliopathic pain: When satellite glial cells go bad." *The Neuroscientist* 15 (5): 450–63.

Palma-Gudiel, Helena, Aric A Prather, Jue Lin, Jake D Oxendine, Jerry Guintivano, Kai Xia, David R Rubinow, Owen Wolkowitz, Elissa S Epel, and Anthony S Zannas. 2021. "HPA axis regulation and epigenetic programming of immune-related genes in chronically stressed and non-stressed mid-life women." *Brain, Behavior, and Immunity* 92: 49–56.

Pierre, Kamau, Naomi Schlesinger, and Ioannis P Androulakis. 2016. "The role of the hypothalamic-pituitary-adrenal axis in modulating seasonal changes in immunity." *Physiological Genomics* 48 (10): 719–38.

Pinho-Ribeiro, Felipe A, Buket Baddal, Rianne Haarsma, Maghnus O'Seaghdha, Nicole J Yang, Kimbria J Blake, Makayla Portley, Waldiceu A Verri, James B Dale, and Michael R Wessels. 2018. "Blocking neuronal signaling to immune cells treats streptococcal invasive infection." *Cell* 173 (5): 1083–97.

Posey, William C, Harold S Nelson, Bernard Branch, and David S Pearlman. 1978. "The effects of acute corticosteroid therapy for asthma on serum immunoglobulin levels." *Journal of Allergy and Clinical Immunology* 62 (6): 340–48.

Potter, Gregory D M, Debra J Skene, Josephine Arendt, Janet E Cade, Peter J Grant, and Laura J Hardie. 2016. "Circadian rhythm and sleep disruption: causes, metabolic consequences, and countermeasures." *Endocrine Reviews* 37 (6): 584–608.

Pradhan, Arpit Kumar, Tatjana Neumüller, Claudia Klug, Severin Fuchs, Martin Schlegel, Markus Ballmann, Katharina Johanna Tartler, et al. 2023. "Chronic administration of XBD173 ameliorates cognitive deficits and neuropathology via 18 KDa Translocator Protein (TSPO) in a mouse model of Alzheimer's Disease." *BioRxiv*, January, 2023.02.23.529740. https://doi.org/10.1101/2023.02.23.529740

Radjavi, Ali, Igor Smirnov, and Jonathan Kipnis. 2014. "Brain antigen-reactive CD4+ T cells are sufficient to support learning behavior in mice with limited T cell repertoire." *Brain, Behavior, and Immunity* 35: 58–63.

Radley, Jason J, and Paul E Sawchenko. 2011. "A common substrate for prefrontal and hippocampal inhibition of the neuroendocrine stress response." *Journal of Neuroscience* 31 (26): 9683–95.

Rainer, Rupprecht, Arpit Kumar Pradhan, Marco Kufner, Lisa Marie Brunner, Caroline Nothdurfter, Simon Wein, Jens Schwarzbach, Xenia Puig, Christian Rupprecht, Gerhard Rammes. 2022. "Neurosteroids, translocator protein 18 KDa (TSPO) in depression: Implications for synaptic plasticity, cognition, and treatment options." *European Archives of Psychiatry and Clinical Neuroscience* In press.

Ransohoff, Richard M, Pia Kivisäkk, and Grahame Kidd. 2003. "Three or more routes for leukocyte migration into the central nervous system." *Nature Reviews Immunology* 3 (7): 569–81.

Reed, David E, Carlos Barajas-Lopez, Graeme Cottrell, Sara Velazquez-Rocha, Olivier Dery, Eileen F Grady, Nigel W Bunnett, and Stephen J Vanner. 2003. "Mast cell tryptase and proteinase-activated receptor 2 induce hyperexcitability of guinea-pig submucosal neurons." *The Journal of Physiology* 547 (2): 531–42.

Rennels, M L, O R Blaumanis, and P A Grady. 1990. "Rapid solute transport throughout the brain via paravascular fluid pathways." *Advances in Neurology* 52: 431–39.

Rennels, Marshall L, Thomas F Gregory, Otis R Blaumanis, Katsukuni Fujimoto, and Patricia A Grady. 1985. "Evidence for a 'Paravascular' fluid circulation in the mammalian central nervous system, provided by the rapid distribution of tracer protein throughout the brain from the subarachnoid space." *Brain Research* 326 (1): 47–63.

Riol-Blanco, Lorena, Jose Ordovas-Montanes, Mario Perro, Elena Naval, Aude Thiriot, David Alvarez, Silke Paust, John N Wood, and Ulrich H von Andrian. 2014. "Nociceptive sensory neurons drive Interleukin-23-Mediated psoriasiform skin inflammation." *Nature* 510 (7503): 157–61.

Romano-Keeler, Joann, and Jörn-Hendrik Weitkamp. 2015. "Maternal influences on fetal microbial colonization and immune development." *Pediatric Research* 77 (1): 189–95.

Rupprecht, Rainer, Vassilios Papadopoulos, Gerhard Rammes, Thomas C Baghai, Jinjiang Fan, Nagaraju Akula, Ghislaine Groyer, David Adams, and Michael Schumacher. 2010. "Translocator Protein (18 KDa)(TSPO) as a therapeutic target for neurological and psychiatric disorders." *Nature Reviews Drug Discovery* 9 (12): 971–88.

Sakakibara, Yasufumi, Michiko Sekiya, Takashi Saito, Takaomi C Saido, and Koichi M Iijima. 2019. "Amyloid-β plaque formation and reactive gliosis are required for induction of cognitive deficits in app knock-in mouse models of Alzheimer's Disease." *Bmc Neuroscience* 20 (1): 1–14.

Schotanus, KAREL, F J Tilders, and FRANK Berkenbosch. 1993. "Human recombinant interleukin-1 receptor antagonist prevents adrenocorticotropin, but not interleukin-6 responses to bacterial endotoxin in rats." *Endocrinology* 133 (6): 2461–68.

Serhan, Nadine, Lilian Basso, Riccardo Sibilano, Camille Petitfils, James Meixiong, Chrystelle Bonnart, Laurent L Reber, Thomas Marichal, Philipp Starkl, and Nicolas Cenac. 2019. "House dust mites activate nociceptor–mast cell clusters to drive Type 2 skin inflammation." *Nature Immunology* 20 (11): 1435–43.

Shanahan, FJJJA, J A Denburg, J Fox, J Bienenstock, and D Befus. 1985. "Mast cell heterogeneity: Effects of neuroenteric peptides on histamine release." *Journal of Immunology (Baltimore, Md.: 1950)* 135 (2): 1331–37.

Shi, Limin, S Hossein Fatemi, Robert W Sidwell, and Paul H Patterson. 2003. "Maternal influenza infection causes marked behavioral and pharmacological changes in the offspring." *Journal of Neuroscience* 23 (1): 297–302.

Skaper, Stephen D. 2007. "The brain as a target for inflammatory processes and neuroprotective strategies." *Annals of the New York Academy of Sciences* 1122 (1): 23–34.

Solomon, Matia B, Matthew Loftspring, Annette D De Kloet, Sriparna Ghosal, Ryan Jankord, Jonathan N Flak, Aynara C Wulsin, Eric G Krause, Rong Zhang, and Taylor Rice. 2015. "Neuroendocrine function after hypothalamic depletion of glucocorticoid receptors in male and female mice." *Endocrinology* 156 (8): 2843–53.

Spichak, Simon, Timothy G Dinan, and John F Cryan. 2019. "Gut–neuroimmune interactions: The unexpected role of the immune system in brain development." *The Biochemist* 41 (1): 36–41.

Steinhoff, M, N Vergnolle, S H Young, M Tognetto, Silvia Amadesi, H S Ennes, M Trevisani, M D Hollenberg, J L Wallace, and G H Caughey. 2000. "Agonists of proteinase-activated receptor 2 induce inflammation by a neurogenic mechanism." *Nature Medicine* 6 (2): 151–58.

Stephens, Mary Ann C, and Gary Wand. 2012. "Stress and the HPA axis: Role of glucocorticoids in alcohol dependence." *Alcohol Research: Current Reviews* 34 (4): 468.

Szatmari, Istvan, and Laszlo Nagy. 2008. "Nuclear receptor signalling in dendritic cells connects lipids, the genome and immune function." *The EMBO Journal* 27 (18): 2353–62.

Talaber, Gergely, Jan Peter Tuckermann, and Sam Okret. 2015. "ACTH controls thymocyte homeostasis independent of glucocorticoids." *The FASEB Journal* 29 (6): 2526–34.

Tamura, Kenji, and Jackie D Wood. 1992. "Effects of prolonged exposure to histamine on guinea pig intestinal neurons." *Digestive Diseases and Sciences* 37: 1084–88.

Thion, Morgane Sonia, Donovan Low, Aymeric Silvin, Jinmiao Chen, Pauline Grisel, Jonas Schulte-Schrepping, Ronnie Blecher, Thomas Ulas, Paola Squarzoni, and Guillaume Hoeffel. 2018. "Microbiome influences prenatal and adult microglia in a sex-specific manner." *Cell* 172 (3): 500–16.

Tsitoura, Daphne C, and Paul B Rothman. 2004. "Enhancement of MEK/ERK signaling promotes glucocorticoid resistance in CD4+ T Cells." *The Journal of Clinical Investigation* 113 (4): 619–27.

Turnbull, Andrew V, and Catherine L Rivier. 1999a. "Regulation of the hypothalamic-pituitary-adrenal axis by cytokines: Actions and mechanisms of action." *Physiological Reviews* 79 (1): 1–71.

Turnbull, Andrew V, and Catherine L Rivier. 1999b. "Sprague-dawley rats obtained from different vendors exhibit distinct adrenocorticotropin responses to inflammatory stimuli." *Neuroendocrinology* 70 (3): 186–95.

Vahl, Torsten P, Yvonne M Ulrich-Lai, Michelle M Ostrander, C Mark Dolgas, Eileen E Elfers, Randy J Seeley, David A D'Alessio, and James P Herman. 2005. "Comparative analysis of ACTH and corticosterone sampling methods in rats." *American Journal of Physiology-Endocrinology and Metabolism* 289 (5): E823–28.

Viblanc, Vincent A, Quentin Schull, Tina Cornioley, Antoine Stier, Jean-Jérôme Ménard, René Groscolas, and Jean Patrice Robin. 2018. "An integrative appraisal of the hormonal and metabolic changes induced by acute stress using king penguins as a model." *General and Comparative Endocrinology* 269: 1–10.

Voet, Sofie, Marco Prinz, and Geert van Loo. 2019. "Microglia in central nervous system inflammation and multiple sclerosis pathology." *Trends in Molecular Medicine* 25 (2): 112–23.

Voisin, Tiphaine, Amélie Bouvier, and Isaac M Chiu. 2017. "Neuro-immune interactions in allergic diseases: Novel targets for therapeutics." *International Immunology* 29 (6): 247–61.

Wang, Qinqin, Hongmei Yao, Wenyan Liu, Bailiu Ya, HongJu Cheng, Zhenkai Xing, and Yili Wu. 2021. "Microglia polarization in Alzheimer's disease: Mechanisms and a potential therapeutic target." *Frontiers in Aging Neuroscience* 13.

Wang, Yaming, Kristy J Szretter, William Vermi, Susan Gilfillan, Cristina Rossini, Marina Cella, Alexander D Barrow, Michael S Diamond, and Marco Colonna. 2012. "IL-34 is a tissue-restricted ligand of CSF1R required for the development of langerhans cells and microglia." *Nature Immunology* 13 (8): 753–60.

Wei, Dereck C, and Elizabeth H Morrison. 2019. "Histology, astrocytes."

Weinstein, Robert S, Robert L Jilka, A Michael Parfitt, and Stavros C Manolagas. 1998. "Inhibition of osteoblastogenesis and promotion of apoptosis of osteoblasts and osteocytes by glucocorticoids. potential mechanisms of their deleterious effects on bone." *The Journal of Clinical Investigation* 102 (2): 274–82.

Weller, Roy O, Effie Djuanda, Hong-Yeen Yow, and Roxana O Carare. 2009. "Lymphatic drainage of the brain and the pathophysiology of neurological disease." *Acta Neuropathologica* 117: 1–14.

Wolf, Susanne A, Barbara Steiner, Akgul Akpinarli, Thomas Kammertoens, Christina Nassenstein, Armin Braun, Thomas Blankenstein, and Gerd Kempermann. 2009. "CD4-Positive T Lymphocytes provide a neuroimmunological link in the control of adult hippocampal neurogenesis." *The Journal of Immunology* 182 (7): 3979–84.

Wu, Hsin-Jung, and Eric Wu. 2012. "The role of gut microbiota in immune homeostasis and autoimmunity." *Gut Microbes* 3 (1): 4–14.

Yi, Chun-Xia, Susanne E la Fleur, Eric Fliers, and Andries Kalsbeek. 2010. "The role of the autonomic nervous liver innervation in the control of energy metabolism." *Biochimica et Biophysica Acta (BBA)-Molecular Basis of Disease* 1802 (4): 416–31.

Zieg, George, Gideon Lack, Ronald J Harbeck, Erwin W Gelfand, and Donald Y M Leung. 1994. "In vivo effects of glucocorticoids on IgE production." *Journal of Allergy and Clinical Immunology* 94 (2): 222–30.

Ziegon, Laura, and Martin Schlegel. 2021. "Netrin-1: A modulator of macrophage driven acute and chronic inflammation." *International Journal of Molecular Sciences* 23 (1): 275.

2 Neuroimmune System Scenario in Neurodegenerative Disorders

Arpit Kumar Pradhan
University of Michigan, Ann Arbor, MI, USA

Mirudula Elanchezhian
Indian Institute of Science Education and Research, Mohali, India

2.1 INTRODUCTION

Neurodegeneration represents a multifaceted group of disorders marked by the gradual and relentless decline in both the structural integrity and functional capacity of neurons within the central nervous system (CNS). This category of diseases encompasses a diverse spectrum of conditions, each with its unique characteristics and challenges. At its core, neurodegeneration involves a progressive and often irreversible deterioration of nerve cells. This deterioration can manifest in various ways, such as cognitive decline, motor dysfunction, and sensory impairments, depending on the specific disorder and the affected neural pathways (Mayne et al. 2020).

Notable examples include Alzheimer's disease, Parkinson's disease, amyotrophic lateral sclerosis (ALS), and Huntington's disease (HD), among numerous others. Despite the wide-ranging clinical presentations, which stem from the loss of specific neurons and synapses in distinct brain regions, these conditions share commonalities in terms of features and underlying mechanisms. One defining feature of these neurodegenerative diseases is the accumulation of abnormal proteins within the brain and nervous system (Taylor, Hardy, and Fischbeck 2002; Gan et al. 2018). These protein aggregates, like amyloid plaques and tau tangles in Alzheimer's disease or Lewy bodies in Parkinson's disease, disrupt the normal functioning of neurons, contributing to their demise. This abnormal protein buildup is a central focus of research into the underlying causes of these conditions.

Though initially it was believed that the nervous and immune systems operated as separate entities, Ader and Cohen's groundbreaking research illustrated that they could intensify the immunosuppressive effects of cyclophosphamide in rats by exposing

DOI: 10.1201/9781003295471-2

them to the taste of a saccharin solution that had earlier been linked to the drug's negative consequences (Ader and Cohen 1975). This experiment offered indirect revelations about the intricate network of bidirectional communication between the CNS and the immune system. Numerous studies have outlined the fundamental modes of communication between the nervous and immune systems. These include the production and functionality of neuroendocrine peptide hormones by the immune system, the impact of neuroendocrine hormones on immune system functioning (Madden and Felten 1995), the sympathetic nervous system's involvement in innervating lymphoid organs, and the regulatory influence of cytokines on the hypothalamic-pituitary-adrenal (HPA) axis (Turnbull and Rivier 1999).

The adaptive immune system plays a significant role in maintaining CNS homeostasis, which refers to the stable and balanced internal environment of the CNS. For many years, it was widely believed that the CNS and the retina were immune-privileged due to their protection by the blood-brain barrier (BBB), blood-cerebrospinal fluid barrier (BCSFB), or the blood-retinal barrier (BRB). The immune response to foreign tissues within the CNS, except for tumor-related situations, is not as swift as it would be in other body regions. Nevertheless, it's crucial to understand that this delayed immune reaction doesn't suggest that the CNS is completely shielded from immune interactions (Galea, Bechmann, and Perry 2007). A diverse array of immune cells, including dendritic cells (DCs), border-associated macrophages (BAMs), T cells, innate lymphoid cells (ILCs), neutrophils, and B cells, plays a pivotal role in monitoring the immune status of the brain. These immune cells are strategically distributed throughout the meninges and other cerebrospinal fluid (CSF)–containing regions of the CNS, such as perivascular spaces and the choroid plexus (Kivisäkk et al. 2003; Jordão et al. 2019; Van Hove et al. 2019).

Meningeal DCs and BAMs serve as antigen-presenting cells (APCs), and they actively participate in the presentation of locally derived CNS antigens to patrolling T cells. This interaction facilitates robust immune surveillance within the brain. Consequently, the meninges emerge as a pivotal interface for communication between the CNS and the immune system, playing a critical role in maintaining the CNS's immune integrity (Rustenhoven et al. 2021). In addition, it has been demonstrated that neurogenesis relies, in part, on the presence of the adaptive immune system. Specifically, a deficiency in T cells has been linked to impaired neurogenesis in the hippocampus, resulting in cognitive deficits (Ziv et al. 2006).

This newfound understanding holds promise for future research and therapeutic strategies aimed at tackling neurodegenerative diseases. As we delve deeper into these complex interactions, we move closer to unlocking potential treatments and interventions for these debilitating conditions.

2.2 ROLE OF GLIAL CELLS AND MACROPHAGES IN BRAIN HOMEOSTASIS AND DISEASE

The entire population of glial cells can be categorized into four primary groups, which include (1) microglia, (2) astrocytes, (3) oligodendrocytes, and (4) their precursor cells known as NG2-glia. Microglia make up approximately 5–10% of the total brain cell population and represent the exclusive macrophages within the CNS

(Aguzzi, Barres, and Bennett 2013). Microglia, along with perivascular, meningeal, and choroid plexus macrophages, each occupy specific strategic niches, collectively ensuring comprehensive coverage of the entire CNS (Prinz, Erny, and Hagemeyer 2017; Goldmann et al. 2016). Astrocytes are the most prevalent cell type in the brain (~90% of overall brain mass) (Kettenmann and Ransom 2004). NG2-glia, part of the oligodendrocyte lineage, continuously generate mature oligodendrocytes (Dimou et al. 2008; Rivers et al. 2008). In the adult rodent brain, they form a tight homeostatic network where cell numbers are maintained. If one cell differentiates or dies, a neighboring cell quickly takes its place (Hughes et al. 2013). The function of mature oligodendrocytes is to produce myelin, insulating axons for rapid conduction and supporting axons (Nave 2010).

Microglia play a role in various brain functions, including developmental processes, homeostasis maintenance, tissue repair, and involvement in disease pathology, by interacting with nearly all types of brain cells (Frost and Schafer 2016; Schafer and Stevens 2015). Increasing evidence suggests that microglia play an active role in synapse remodeling and maturation during development (Hoshiko et al., 2012; Paolicelli et al., 2011; Schafer et al., 2012). For instance, Paolicelli and colleagues demonstrated an unconventional role of microglia in facilitating the maturation and plasticity of hippocampal synapses, potentially by eliminating immature synapses (Paolicelli et al., 2011). Microglia in the brain are key producers of vital complement components like C1q, CR3, and CR5 (Y. Zhang et al., 2014; Stephan, Barres, and Stevens 2012). These complement proteins have diverse functions beyond their traditional immune roles and are involved in developmental synapse pruning, disease-related synapse loss, and age-related cognitive decline (Stephan, Barres, and Stevens 2012). During the first postnatal week, microglia help refine synapses in the visual system, tagging weaker ones with C1q and C3 and engulfing them via CR3. This process isn't limited to the visual system but occurs throughout the developing CNS, suggesting a broader role for C1q tagging in microglial synapse pruning (Schafer et al., 2012; B. Stevens et al., 2007).

Microglia are involved in regulating the neuronal population by engulfing neural precursor cells (NPCs) during embryonic and adult neurogenesis (Cunningham, Martínez-Cerdeño, and Noctor 2013; Sierra et al., 2010). Factors like maternal immune challenges or inhibiting microglial activation can influence this process. Microglia use TAM receptors: TYRO3, AXL, and MER receptor tyrosine kinases and their ligands, including growth-arrest-specific protein 6 (GAS6) and vitamin K-dependent protein S, for this phagocytic function (Fourgeaud et al., 2016). Interestingly, elevated AXL expression is observed in several neurodegenerative disease models, suggesting a potential link between abnormal microglial phagocytosis and disease pathology (Chiu et al., 2013; Yin et al., 2017).

Astrocytes maintain brain homeostasis by regulating K+ levels, neurotransmitters, and the BBB, and supplying nutrients. They store glycogen, produce ApoE for brain health, and modulate immune responses (Gee and Keller 2005). Astrocytes can function as antigen-presenting cells (APCs), producing major histocompatibility complex (MHC) class I and II molecules and intercellular adhesion molecules (ICAMs). They can affect BBB permeability, allowing T cells to enter the brain and modulate the immune response with cytokine release. Astrocytes also produce various cytokines, including interleukin (IL), tumor necrosis factor (TNF), and interferon (IFN) (Gee and Keller 2005). Astrocytes contribute to age-related diseases and normal aging changes, impacting synaptic plasticity and inflammation (Nagele et al. 2004; Teismann et al. 2003). Their ability to produce growth factors and cytokines suggests that age-related alterations in astrocyte function may affect neuronal survival (Cotrina and Nedergaard 2002).

Following injury, astrocytes undergo a process referred to as "astrogliosis," which involves significant changes in both morphology and function. Astrogliosis encompasses a wide range of reactivity, characterized by the increased expression of structural proteins like glial fibrillary acidic protein (GFAP) and vimentin, as well as the enlargement of both the cell body and processes (Wilhelmsson et al., 2004; Sofroniew and Vinters 2010). Similar to astrocytes and microglia, NG2-glia have been observed to respond to different types of pathological insults, but only when the BBB is compromised (A. Wang and He 2009; Rhodes, Raivich, and Fawcett 2006; Dimou and Götz 2014). NG2-glia significantly proliferate and form a perimeter around the injury site as part of the glial scar (Levine, Reynolds, and Fawcett 2001). However, in repeated in vivo imaging, it has been observed that, unlike microglia and NG2-glia, astrocytes are selective of migrating into the injury core following mechanical injury (Bardehle et al., 2013).

Microglia engage in interactions with astrocytes, particularly in situations of brain injury and inflammation. When an injury triggers the release of ATP, astrocytes produce an ATP gradient. Microglia detect this gradient through their purinergic receptor P2RY12, which prompts the rapid convergence of microglial processes and cell migration (Davalos et al., 2005; Haynes et al., 2006). In response to injury or LPS stimulation, activated microglia release IL-1α, TNF, and C1q. These molecules are sufficient to trigger the formation of 'A1' neurotoxic reactive astrocytes. A1 astrocytes, which are prevalent after CNS injuries and in neurodegenerative diseases, lose their normal synapse-supporting and phagocytic functions while gaining the ability to kill neurons and oligodendrocytes (Liddelow et al., 2017). Pharmacological and genetic studies have demonstrated that during early postnatal development, microglia play a crucial role in the formation of oligodendrocyte progenitors and the subsequent myelination process. In adulthood, microglia continue to support the maintenance of the oligodendrocyte progenitor pool under normal conditions (Hagemeyer et al., 2017).

2.3 NEUROIMMUNE COMMUNICATIONS AND ALZHEIMER'S DISEASE

Alzheimer's disease (AD) is a complex and devastating neurodegenerative disorder that primarily affects cognitive function, memory, and behavior. While the hallmark pathological features of AD include the accumulation of β-amyloid (Aβ) plaques and tau tangles in the brain, there is growing recognition that the immune system and neuroinflammatory processes play pivotal roles in the disease's development and progression (Jack Jr et al., 2018).

The intricate crosstalk between the nervous system and the immune system, often referred to as neuroimmune communication, is of paramount importance in understanding AD. This communication involves various immune cells, cytokines, and other signaling molecules that influence the brain's immune response and, consequently, the course of AD. The initial indications of the involvement of the adaptive immune system in AD were rooted in the observation of elevated CD4 and CD8 T cell counts within the brains of AD patients (Itagaki, McGeer, and Akiyama 1988; Togo et al., 2002).

Further investigations, including studies conducted in mouse models exhibiting β-amyloidosis and analyses of post-mortem brain samples from AD patients, revealed that while micro-hemorrhages contributed to non-specific T cell movement into brain tissues, the specific infiltration of T cells into the brain parenchyma was predominantly associated with the presence of β-amyloidosis and tau pathology (Ferretti et al., 2016; Laurent et al., 2017; Merlini et al., 2018). However, it's worth noting that these T cells were typically not found close to β-amyloid plaques and neurofibrillary tangles (NFTs), suggesting that while the accumulation of cerebral β-amyloid and tau plays a role, it may not be the sole factor driving T cell invasion into the brain (Togo et al., 2002; Laurent et al., 2017; Merlini et al., 2018).

Significantly, the immune receptor TREM2 plays a crucial role in orchestrating microglial responses, which encompass their accumulation near Aβ deposits and the process of phagocytosing Aβ plaques (Ulrich et al., 2014; Yeh et al., 2016). In a study by Wang and colleagues, it was revealed that the administration of an agonist antibody targeting TREM2 promotes the activation of microglia and stimulates the phagocytosis of Aβ in mouse models characterized by β-amyloidosis. This intervention effectively mitigates the toxicity associated with Aβ plaques and reduces neuronal damage (S. Wang et al., 2020b). Antigen-dependent T cell infiltration in the brain during AD necessitates successful antigen presentation and T cell reactivation at the CNS borders. Microglia, located near Aβ plaques and involved in Aβ aggregate phagocytosis, may serve as local APCs for Aβ-reactive T cell reactivation (Perlmutter et al., 1992; Monsonego and Weiner 2003; C. Y. D. Lee and Landreth 2010; Mittal et al., 2019). However, phagocytosis and antigen presentation are compromised in AD (Gu et al., 2016), with Aβ oligomers inhibiting antigen presentation by reducing MHC-II surface expression on APCs and T cell activation in mice with β-amyloidosis (Gericke et al., 2020).

Besides, microglia exhibit the capability to adhere to soluble Aβ oligomers and Aβ fibrils through the engagement of receptors. These receptors encompass class A scavenger receptor A1, CD36, CD14, α6β1 integrin, CD47, as well as toll-like receptors (TLR2, TLR4, TLR6, and TLR9) (Bamberger et al., 2003; Paresce, Ghosh, and Maxfield 1996; Stewart et al., 2010; Y. Liu et al., 2005). This binding is believed to be a component of the inflammatory response associated with AD. The activation of microglia is both defined and regulated by cytokines. In aging transgenic mice like TgAPPsw and PSAPP, the elevated levels of Aβ are linked to heightened levels of pro-inflammatory cytokines, such as TNF-α, IL-6, IL-1α, and granulocyte macrophage-colony stimulating factor (GM-CSF) (Patel et al., 2005).

Similar to microglia, astrocytes also release cytokines, ILs, nitric oxide, and other potentially harmful molecules when exposed to Aβ, thereby contributing to the escalation of neuroinflammation. The significance of astroglial inflammation in AD was recently explored through AAV-driven suppression of the astrocytic response in APP/PS1 mice. This intervention, which disrupted the calcineurin/nuclear factor of the activated T-cells signaling pathway, resulted in improved cognitive function, reduced astrogliosis, and lower Aβ levels (Furman et al., 2012). While, the astrocytes mediate synapse elimination in a C1q-independent fashion in a physiological state, in pathological scenarios astrocytes perform synapse elimination in a C1q-dependent manner (Dejanovic et al., 2022). Our recent study which targeted the translocator protein via the ligand XBD173 found that it provided neuroprotective benefits by reduced astrocytic elimination of functional synapse in a complement-mediated pathway (Pradhan et al., 2023).

Conversely, it appears that triggering certain pro-inflammatory signaling pathways can have a beneficial impact in mouse models of AD. For instance, in the APP/PS1 mouse model, the transgenic expression of IL-1β resulted in substantial neuroinflammation and a reduction in amyloid plaque pathology (Ghosh et al., 2013; Shaftel et al., 2007). In another investigation involving the TgCRND8 mouse model, the adeno-associated virus (AAV)-mediated expression of IFN-γ in the brain illustrated how this pro-inflammatory cytokine could enhance the clearance of amyloid plaques, albeit accompanied by a widespread increase in astrogliosis and microgliosis (Chakrabarty et al., 2010).

Chemokines in AD have been proposed to regulate the migration of microglia to sites of neuroinflammation, thereby intensifying local inflammation (Savarin-Vuaillat and Ransohoff 2007). In vitro studies have demonstrated that Aβ triggers the production of CXCL8 (IL-8), CCL2, CCL3, and CCL4 in human macrophages and astrocytes (Smits et al., 2002). Additionally, microglia cultured from autopsied AD brains displayed increased expression of CXCL8, CCL2, and CCL3 following experimental exposure to Aβ (Lue, Walker, and Rogers 2001). In AD mouse models, the CX3CR1/CX3CL1 system has been shown to modulate neuronal survival, plaque load, and cognitive function (Fuhrmann et al., 2010; Cho et al., 2011; S. Lee et al., 2010b). Microglia are often considered important targets in AD pathogenesis because they actively participate in functional synapse removal. Microglia are a highly dynamic entity, and previous reports suggest that they can possess both pro- and anti-inflammatory markers (Soldano et al., 2018). However, in AD, there is a shift in microglia responsiveness from anti- to pro-inflammatory markers (Guo, Wang, and Yin 2022; Q. Wang et al., 2021). For instance, microglia have been demonstrated to display inflammatory markers such as CD14, CD11c, CD36, MHCII, and inducible nitric oxide synthase (iNOS) in AD (Tolar et al., 2021; Kamphuis et al., 2016; Hansen, Hanson, and Sheng 2018). Our recent study shows a reduction in the percentage of resting microglia in the transgenic Arc Aβ mice (AD model). Additionally, our *ex vivo* hippocampal slice experiments showed an increase in microglia marker P2RY12 as well as the inflammatory marker CD80 in the transgenic AD mice model (Pradhan et al., 2023).

Caspases are intracellular proteases that regulate apoptosis and inflammation. Caspase-1 is controlled by inflammasomes, activating IL-1β and IL-18 (Van de Veerdonk et al., 2011; Schroder and Tschopp 2010). High caspase-1 levels are found in AD patients and APP/PS1 mice. Removing NLRP3 or caspase-1 in these mice protects against AD-related memory and synaptic issues (Heneka et al., 2013). iNOS levels are also increased in AD brains, and genetic knockout of iNOS is protective in AD mouse models (Nathan et al., 2005; Vodovotz et al., 1996). Additionally, microglia highly express NADPH oxidase (PHOX), which is upregulated in AD and rapidly activated by inflammatory stimuli such as Aβ, leading to the generation of hydrogen peroxide that promotes microglial activation (S. Choi et al., 2012; Jekabsone et al., 2006). Further, there is an association between activated complement factor products and the presence of Aβ deposits (Strohmeyer et al., 2002).

Moreover, the successful discovery of human autoantibodies against Aβ, generated by immortalized B cells from an AD patient, confirmed the effective priming of B cells against aggregated Aβ species in humans (Gaskin et al., 1993). The existence of a potential neuroprotective role for these autoantibodies and their association with the progression of AD have been topics of ongoing debate (Hyman et al., 2001; Britschgi et al., 2009; Dodel et al., 2011; Mengel et al., 2013; Söllvander et al., 2015). Evidence of the detrimental role of lymphocytes in AD was established through research. One study demonstrated that in a β-amyloidosis mouse model (PSAPP, Rag2–/–) lacking functional T and B cells, there was a notable reduction in Aβ pathology (Späni et al., 2015). However, another study, which employed a distinct mouse model (5xfAD, Rag2–/– Il2rg–/–) characterized by the absence of T cells, B cells, and NK cells, revealed an opposite outcome with accelerated Aβ pathogenesis and heightened neuroinflammation (Späni et al., 2015).

Various immunotherapies have been developed to combat AD by targeting Aβ plaques. Active Aβ immunization in mouse models has demonstrated effectiveness in reducing plaque formation and improving memory (Janus et al., 2000; Morgan et al., 2000; Schenk et al., 1999). Additionally, passive immunotherapy with anti-Aβ antibodies, particularly those that target aggregated Aβ, has shown promise in reducing plaque load and slowing cognitive decline. These findings suggest that harnessing humoral immune responses may hold the key to future therapies aimed at mitigating the Aβ burden and ameliorating AD symptoms (Mintun et al., 2021; Sevigny et al., 2016).

2.4 PATHOPHYSIOLOGY AND INVOLVEMENT OF THE IMMUNE SYSTEM IN MULTIPLE SCLEROSIS

Multiple sclerosis (MS) is a chronic inflammatory neurological condition that impacts millions globally. It is an autoimmune disorder that targets the CNS, encompassing the brain, spinal cord, and optic nerves. The disease is marked by focal lymphocytic infiltrates, the breakdown of myelin sheaths surrounding axons, astrogliosis, microglial activation, and widespread neurodegeneration (Hauser 1994). From a pathological perspective, perivascular lymphocytic infiltrates and macrophages are responsible for the degradation of myelin sheaths encasing neurons. Lesions in the CNS associated with MS can manifest at various points in time and appear in different locations within the CNS. This pattern of lesion development is often described as being "disseminated in time and space." The neurological symptoms of MS are diverse and encompass issues like visual disturbances, sensations of numbness and tingling, localized weakness, difficulties with bladder and bowel control, and cognitive impairments. The specific symptoms experienced can vary depending on the location of the lesions within the CNS (Dendrou, Fugger, and Friese 2015).

MS affects approximately 400,000 individuals in the United States and around 2.5 million people worldwide. It is more prevalent in females, with a three-fold higher occurrence compared to males (Dilokthornsakul et al., 2016). Typically, the disease manifests between the ages of 20 and 40, but it can emerge at any age, with nearly 10% of cases appearing before the age of 18 (Goldenberg, 2012). In populations of European ancestry, the overall prevalence is estimated to be 1 in 1000 (Didonna and Oksenberg 2017). Prevalence rates in non-European populations, particularly East Asian and African descent, are not as well-documented, although recent studies have identified a higher prevalence among African-American populations, similar to those of European ancestry (Wallin et al., 2012).

The precise cause of MS remains elusive. Extensive research suggests that a combination of genetic factors and environmental influences plays a multifaceted role in the onset of the condition (Holmøy and Hestvik 2008; Gourraud et al., 2012). In the latter half of the 19th century, the observation of family clustering of MS cases provided initial insights into the genetic aspect of the disease. While a family recurrence rate of about 20%-33%, first-degree relatives, faced a notable 10- to 12-fold increase in risk, while second-degree relatives exhibited a threefold risk elevation (Carton et al., 1997; Robertson et al., 1996). Siblings or children of individuals with MS were estimated to have an overall risk of 3% to 5% (Sadovnick et al., 1998). However, twin studies indicated a higher risk, particularly when the sibling was a monozygotic twin, where the risk reached 25% (Mumford et al., 1994; Willer et al., 2003). Conversely, spouses and adopted individuals maintain a risk similar to that of the general population or their biological families, supporting the notion that shared genetics underlie the familial clustering of MS (Robertson et al., 1996). Indeed, extensive genome-wide association studies have pinpointed over 200 gene variants associated with an increased risk of developing MS (Thompson, Baranzini, et al., 2018b).

The major genetic association in MS is situated within the MHC on chromosome 6p21.3. The initial connection between the human leukocyte antigen (HLA) system and MS risk was established in 1972, when certain surface glycoproteins linked to HLA-A3 and HLA-B7 class I alleles were found more prevalent in MS patients (Naito et al., 1972; Jersild, Svejgaard, and Fog 1972). Subsequent research, regardless of sample size or resolution, has consistently confirmed the association between HLA and MS risk across all populations and MS subtypes. While the initial link was with class I HLA-A and HLA-B alleles, more powerful studies, including genome-wide association studies (GWAS), have identified the primary genetic susceptibility signal for MS as the HLA-DRB1 locus within the class II region of the MHC. The HLA-DRB1*15:01 allele has the strongest effect, with an average odds ratio (OR) of 3.08, and its impact increases with the number of copies an individual carries (Hollenbach et al., 2016; "Genetic Risk and a Primary Role for Cell-Mediated Immune Mechanisms in Multiple Sclerosis" 2011).

Using GWAS single nucleotide polymorphism (SNP) data, the International Multiple Sclerosis Genetics Consortium (IMSGC) identified 11 statistically independent effects in the MHC region. These include six HLA-DRB1 and one HLA-DPB1 allele in the centromeric class II region, one HLA-A and two HLA-B alleles in the telomeric class I region, and one in the class III region between MHC class I chain-related protein B (MICB) and LST1 (Patsopoulos et al., 2013). Further analysis of high density MHC region SNP data in European cohorts revealed both new and previously identified HLA class II risk alleles (e.g., DRB1*15:01) and independent HLA class I protective alleles (e.g., A*02:01), along with evidence of interactions between pairs of class II alleles, like DQA1*01:01–DRB1*15:01 and DQB1*03:01–DQB1*03:02 ("Class II HLA Interactions Modulate Genetic Risk for Multiple Sclerosis" 2015). With the assistance of a large GWAS consortium, researchers have identified over 50 susceptible loci, many of which are linked to the immune system. These loci include genes related to interleukin receptors (IL-2 and IL-7), cytokine pathways (CXCR5, IL-12A, IL-12β, IL-12Rβ1, and TNFα), and co-stimulatory molecules (CD80, CD86, CD37) ("Genetic Risk and a Primary Role for Cell-Mediated Immune Mechanisms in Multiple Sclerosis" 2011).

However, it's important to note that the risk does not reach 100% even in identical twins, suggesting the involvement of additional factors beyond genetic makeup in the development of conditions conducive to the dysregulation of the immune response seen in MS. This broader category of determinants encompasses various elements, including environmental exposures as well as epigenetic influences (including DNA methylation patterns, histone modifications, and non-coding RNAs) (Olsson, Barcellos, and Alfredsson 2017). For instance, a hypothesis suggests that vitamin D3 deficiency could increase the risk of MS. Lower vitamin D3 levels are associated with increased MS risk, often linked to reduced sunlight exposure at higher latitudes (Ascherio and Munger 2007). Further, Epstein-Barr virus (EBV) infection is strongly associated with MS (Ascherio and Munger 2007; Hestvik 2010). Those who had seronegative for EBV have a very low MS risk compared to those who have been infected. A history of infectious mononucleosis and rising EBV antibody levels increase MS risk (Ascherio et al., 2001). EBV might mimic myelin basic protein (MBP) antigens pathogenic antigens by presentation on HLA-DRB1*15:01, thus, connecting environmental and genetic factors (Lang et al., 2002).

An uncontrolled immune response contributes to CNS damage in MS, regardless of the initial trigger. A key aspect of MS lesions is the disruption of the BBB (Engelhardt 2008). Mast cells play a role in releasing histamine and tryptase, which have an impact on the integrity of the BBB and the attraction of inflammatory cells into the CNS. Additionally, the production of tryptase and chymase by mast cells triggers a series of events that activate precursor forms of matrix metalloproteinases (MMPs), contributing to processes associated with neurodegeneration (Baranzini et al., 2000; Sospedra and Martin 2005).

MS involves multiple activation states of microglia and macrophages (Prinz, Jung, and Priller 2019), with some subsets producing neurotoxic substances such as TNF-α, reactive oxygen species, and reactive nitrogen species (including nitric oxide). These neurotoxic products can directly harm the CNS in MS (Raivich and Banati 2004; Yamasaki et al., 2014). Further, astrocytes can contribute to innate inflammation and neurodegeneration in multiple sclerosis (MS) by producing cytokines (like IL-6 and TNF), chemokines (such as CCL2), and neurotoxic substances (like NO). On the other hand, specific subsets of astrocytes can help restrain T cell autoimmunity in the CNS by releasing cytokines (e.g., IL-27) or triggering apoptosis through tumor necrosis factor-related apoptosis-inducing ligand (TRAIL) (Wheeler and Quintana 2019; Quintana 2019). Importantly, astrocytes, microglia, and monocytes don't act in isolation but have complex interactions that can exacerbate CNS pathology and potentially provide new targets for therapeutic strategies (Linnerbauer, Wheeler, and Quintana 2020).

CD4+ T cells play a significant role in MS, with the HLA-DRB1*15:01 haplotype strongly associated with MS risk (J. Wang et al., 2020a). Proinflammatory T helper type 1 (Th1) and Th17 cells, linked to MS pathology (D. Hu et al., 2017), are produced when CD4+ T cells are activated in the presence of specific cytokines. Th1 cells express Tbet and produce IFN-γ when activated with IL-12, while Th17 cells, characterized by RORγt expression and IL-17 production, arise from activation with TGF-β1, IL-23, IL-6, or IL-21 (Grigoriadis and Van Pesch 2015; Quintana, Pérez-Sánchez, and Farez 2014). Conversely, CD4+ regulatory T cells (Tregs) expressing the FoxP3 transcription factor and IL-10-producing CD4+ type 1 regulatory T cells (Tr1) help limit pathogenic T cell responses through various mechanisms, including inhibitory cytokine production, cell contact-dependent cytolysis, and metabolic disruption (Venken et al., 2008; Martinez-Forero et al., 2008).

The significance of CD8+ T cells in MS is evident from two observations: they are more numerous than CD4+ T cells in MS-affected areas (Frischer et al., 2009), and genetic associations in the HLA class I region indicate both protective and risk-increasing alleles (Patsopoulos et al., 2013; Fogdell-Hahn et al., 2000). CD8+ T cells contribute to CNS damage through the release of granzymes and perforins, Fas/FasL-mediated cytotoxicity, secretion of proinflammatory cytokines like IL-17 and lymphotoxin, and production of chemoattractants for CD4+ T cells (Medana et al., 2000; Giuliani et al., 2003; Grigoriadis and Van Pesch 2015; Biddison et al., 1998). Mucosal-associated invariant T cells (MAIT cells), a type of CD8+ T cell found in mucosal areas, are elevated in the MS brain and produce IL-17 (Annibali et al., 2011; Held et al., 2015). Depletion of MAIT cells leads to positive clinical outcomes, indicating their significant role in the disease's development (Abrahamsson et al., 2013).

For over 70 years, the presence of antibodies in the CSF and the detection of oligoclonal bands have been observed in MS patients. When B cells are activated, they differentiate into memory or plasma cells, which produce autoantibodies contributing to antibody-dependent cytotoxicity (ADCC) and complement-induced damage (Bar-Or 2005). These antibody bands are present in over 90% of MS patients, aiding in diagnosis (Thompson, Banwell, et al., 2018a). Apart from antibody production, B cells also release proinflammatory cytokines like lymphotoxin and TNF-α and serve as APCs, playing a role in T cell activation (Bar-Or et al., 2010).

A subset of γδ T cells expressing Vγ2 is present in the CSF of MS patients (Triebel and Hercend 1989). These γδ T cells contribute to demyelination by exhibiting cytotoxic activity against oligodendrocytes through mechanisms like Fas ligand, ADCC, perforin, and granzyme B activation. Additionally, oligodendrocytes, which express hsp60, may stimulate the expansion of γδ T cells, suggesting a reciprocal interaction where oligodendrocytes could be targeted by γδ T cells and also trigger their expansion (Grigoriadis and Van Pesch 2015; Zarobkiewicz et al., 2019).

CD56 NK cells are found at reduced levels in individuals with MS (Munschauer et al., 1995) and appear to exert an immunoregulatory influence on the disease. They achieve this through the secretion of immunomodulatory cytokines such as IL-5, IL-13, and TGF-β, as well as by inducing the lysis of target cells using mechanisms involving perforin and TRAIL (Matsumoto et al., 1998; B. Zhang et al., 1997). Furthermore, recent research has revealed that NK cells can stimulate astrocytes to express TRAIL, which, in turn, can promote the apoptosis of T cells driven by

TRAIL. Notably, a study on the therapeutic effects of monoclonal antibodies targeting IL-2 (daclizumab) in MS demonstrated an expansion of CD56 NK cells alongside significant disease improvement (Bielekova et al., 2006).

Current MS treatments aim to modulate the immune system for both symptom management and disease progression control. Broad immunomodulatory drugs, like β-IFN, glatiramer acetate, dimethyl fumarate, and diroximel fumarate, regulate immune responses and reduce inflammation. Sphingosine-1-phosphate receptor modulators (e.g., fingolimod) and natalizumab prevent immune cell infiltration into the central nervous system, reducing relapses. Therapies like ocrelizumab and alemtuzumab deplete specific lymphocytes involved in MS. Antigen-specific approaches, using autologous dendritic cells, TCR peptide vaccines, and recombinant antibodies targeting DEC-205, focus on selectively altering immune responses against encephalitogenic T cells, potentially minimizing side effects. These promising treatments offer more selective and effective immunomodulation, improving MS management (Bourque and Hawiger 2021).

2.5 INFLAMMATION AND NEUROIMMUNE COMMUNICATIONS IN PARKINSON'S DISEASE

Parkinson's disease (PD) is a progressive neurodegenerative condition characterized by the degeneration of dopaminergic neurons in the substantia nigra (SN) and the presence of abnormal protein clusters known as Lewy bodies (Lees, Hardy, and Revesz 2009). The second most prevalent neurodegenerative disorder is PD, and it is projected to double in incidence by 2040, primarily due to the global aging population (Kalia and Lang 2015; Dorsey et al., 2018). While historically viewed primarily as a movement disorder with motor symptoms like resting tremor, bradykinesia, postural instability, and limb rigidity, our understanding of PD has evolved (Dickson 2012). We now recognize it as a multi-faceted disorder, encompassing non-motor symptoms. These non-motor symptoms include sleep disturbances and gastrointestinal issues (Postuma et al. 2012). GWAS suggests that the dysregulation of innate and adaptive immunities may be a key contributor to PD pathogenesis (Pierce and Coetzee 2017).

In PD, inflammation and immune dysfunction are closely linked. Changes in inflammation markers and immune cells can initiate and worsen neuroinflammation, impacting the disease's progression (Williams et al., 2022). Neuroinflammation in the substantia nigra pars compacta (SNpc) due to astrocyte reactivity is a significant factor in PD. The expression of PD-related genes in both neurons and glial cells indicates that mutated gene products in microglia and astrocytes may contribute to PD advancement (J.-Y. Wang et al., 2016; F. W. Booth et al., 2017a). Gradual activation of microglial cells, characterized by markers like MHC-II, ICAM-1, and lymphocyte function-associated antigen 1 (LFA-1), is evident in the SN of PD patients (Rocha, De Miranda, and Teixeira 2015). The extent of microglial activation is associated with early PD's loss of dopaminergic terminals (Ouchi et al., 2005). Microglial profiles in the midbrain, particularly the SNpc, differ from those in other areas of the CNS. This distinction may contribute to the specific vulnerability of dopamine neurons in PD (Surmeier, Obeso, and Halliday 2017). Research has shown that SNpc neurons have unique transcriptional, morphological, and functional characteristics compared to neurons in the ventral tegmental area (VTA), another dopaminergic region that experiences less degeneration in PD (De Biase et al., 2017).

Microglia also help clear necrotic and apoptotic cells and remove toxic protein aggregates, like α-synuclein (Yuan et al., 2022). The TREM-2 receptor variant (p.R47H) expressed on myeloid cells-2 is linked to PD (Graeber 2010). In PD's inflammatory process, TREM2 affects microglial activation and autophagy through the mTOR/p38 MAPK pathway, contributing to PD pathology (Huang et al., 2021). Positron emission tomography (PET) studies show reactive microglia in toxin-induced and transgenic PD mouse models (Bartels et al., 2010). When mice are exposed to human α-synuclein aggregates, α-synuclein interacts with microglia, mediated by Fc-gamma receptors (FcγRs), and is transported to autophagosomes. This interaction affects NF-κB signaling and intracellular trafficking (Cao, Standaert, and Harms 2012). A recent study found that the uptake of α-synuclein fibrils by microglia is regulated by Fyn kinase and the class B scavenger receptor CD36 (Panicker et al., 2019).

In addition, the neurotoxin 1-Methyl-4-phenyl-1,2,3,6-tetrahydropyridine (MPTP) is known to harm dopaminergic neurons by inducing mitochondrial dysfunction and indirectly activating microglia (Dauer and Przedborski 2003; Yu Tang and Le 2016). Other studies found that α-synuclein induces MMPs in primary cultured microglia (E.-J. Lee et al., 2010a). Overexpressing human A53T α-synuclein in mice without the fractalkine receptor (CX3CR1) increased neurodegeneration and neuroinflammation in the SN (Castro-Sánchez et al., 2018). In marmosets, overexpression of wild-type or A53T α-synuclein led to long-lasting changes in microglial morphology, indicating prolonged microglial involvement in α-synucleinopathy (Barkholt et al., 2012). Jmjd3, a protein, is crucial for the development of the neuroprotective M2 microglial phenotype. Inhibiting Jmjd3 reduces neuroprotective polarization and increases pro-inflammatory microglial response, leading to worsened dopaminergic neuron cell death in a PD mouse model (Y Tang et al., 2014).

In some developmental stages or during chronic neuroinflammation, peripheral macrophages may express common microglia markers (Grassivaro et al., 2020). In PD patients, niacin reduces GPR109a expression in macrophages, promoting anti-inflammatory factor secretion, and slowing PD progression (Han et al., 2019; Wakade et al., 2018). Mice injected with MPTP show increased peripheral macrophage levels (Yan et al., 2018). Macrophages can activate PD-related genes, including LRRK2, through pathogen- or sterile-induced endomembrane damage (Herbst et al., 2020). Additionally, glial-derived neurotrophic factor (GDNF) delivery via bone marrow macrophages improves neuroinflammation and inhibits dopaminergic neurodegeneration (C. Chen et al., 2018a).

The disruption of the BBB in PD patients (Gray and Woulfe 2015; Cabezas et al., 2014), coupled with the known roles of GDNF in promoting dopaminergic neuron survival and differentiation (Lf 1993), underscores the significance of astrocyte dysfunction in PD onset and progression. While normally functioning astrocytes play a supportive role, their loss-of-function contributes to the disease. Conversely, reactive astrocytes triggered by CNS stimuli or injuries promote PD pathogenesis through a toxic gain-of-function mechanism (Sofroniew 2009; Phatnani and Maniatis 2015).

These reactive astrocytes release various neurotoxic chemokines and cytokines, including TNF-α and IL-1β (Leal et al., 2013; Lau and Yu 2001). The protein deglycase DJ-1 regulates astrocyte activation through IFN-γ and TLR4 signaling (J. Kim et al., 2013; Oksanen et al., 2019). DJ-1 also plays a role in regulating the inflammatory response in astrocytes. When DJ-1 is absent (DJ-1−/−) in astrocytes, it leads to the production of proinflammatory molecules, including cyclooxygenase-2 (COX-2) and IL-6. This inflammatory response can result in neuronal death, especially when astrocytes are exposed to LPS treatment. DJ-1 appears to selectively affect the TLR4 pathway, but not the TLR3 pathway, in modulating astrocyte inflammation (Waak et al., 2009).

A recent investigation has revealed that activated microglia can trigger the formation of neurotoxic A1 astrocytes by releasing IL-1α, TNF-α, and C1q. Importantly, postmortem brain analyses of individuals with neurodegenerative diseases, including PD, have identified the presence of A1 astrocytes (Liddelow et al., 2017). Genes associated with both PD and astrocyte biology, such as PARK7, SNCA, PARK2, PLA2G6, ATP13A2, LRRK2, GBA, and PINK1, play a role in this process (H. D. E. Booth, Hirst, and Wade-Martins 2017b). For instance, in PD progression, α-synuclein accumulates in neurons, often due to SNCA gene disruptions. Notably, astrocytes have lower SNCA expression than neurons (Solano et al., 2000). Additionally, it has been observed that pathological α-synuclein contributes to the development of A1 astrocytes. In a mouse model of sporadic PD, inhibiting α-synuclein-induced microglial activation and the conversion of astrocytes into the A1 subtype has been shown to protect against dopaminergic neurodegeneration and related behavioral deficits (Yun et al., 2018).

In a study involving induced pluripotent stem cell (iPSC)-derived astrocytes with the familial LRRK2 G2019S mutation, researchers observed dysfunctional chaperone-mediated autophagy (CMA), impaired macroautophagy, and a progressive accumulation of α-synuclein in astrocytes associated with PD (di Domenico et al., 2019). Another recent study, using a knock-in mouse model with reduced phosphorylation of LRKK2 at S910 and S935, observed similar findings in both cell culture models and postmortem brain tissues of PD patients (Nichols et al., 2010; Dzamko et al., 2017). Notably, in these studies, the striatal pathology in PD was linked to a reduced number of astrocytes, while the number of microglia remained unchanged (Zhao et al., 2018). A study revealed that primary astrocytes deficient in GBA exhibit defective autophagic and proteasomal machinery. Furthermore, GBA knockout (KO) neurons and astrocytes displayed mitochondrial dysfunction, including impaired respiration, reduced membrane potential, and more pronounced mitochondrial fragmentation in astrocytes compared to neurons (Osellame and Duchen 2013; Osellame et al., 2013). Further, extensively studied inflammasomes in PD belong to the NLRP3 family, including ASC and caspase-1. When misfolded α-synuclein accumulates, persistent NLRP3 inflammasome activation leads to dopaminergic neuronal cell death due to proinflammatory cytokines. Inflammasomes contribute to the disease's progression and the spread of protein aggregates in the CNS through chronic activation (Ising et al., 2019). Maintaining proper levels of ATP13A2 can help prevent NLPR3 inflammasome activation (H. D. E. Booth, Hirst, and Wade-Martins 2017b; Oksanen et al., 2019).

In addition, monocytes also contribute to disease pathogenesis. In PD, classical CD14+CD16– monocyte frequencies rise, linked to altered transcriptomes (Grozdanov et al., 2014). Higher CCL2 suggests increased monocyte recruitment and inflammation. Elevated LRRK2 levels contribute to monocyte dysfunction in PD (Bliederhaeuser et al., 2016; Cook et al., 2017). Early-stage patients exhibit distinct gene expression patterns, including immune activation genes like HLA-DQB1, MYD88, REL, and TNF (Schlachetzki et al., 2018). A recent transcriptome-wide study positively identified gene links between lysosomal pathways, innate immune function, and PD risk in the dorsolateral prefrontal cortex and peripheral monocytes (J. Li et al., 2019a). DJ-1 deficiency, a PD-related gene, reduced monocyte brain infiltration, delaying injury repair in mice (D. Choi et al., 2020). In the acute MPTP model, deleting Ccr2, vital for monocyte tissue infiltration, had no impact on neurodegeneration. Conversely, increasing monocyte infiltration by overexpressing the Ccr2 ligand-induced neurotoxicity (Parillaud et al., 2017). Intriguingly, in the chronic MPTP model, CNS-infiltrating monocytes proved neurotoxic (L. Gao et al., 2015).

Moreover, functional changes in peripheral system DCs in PD patients have been investigated. Circulating myeloid DCs (MDCs) and lymphocyte-like DCs (LDCs) alteration were detected in PD patient sera. MDCs migrate to inflamed tissues and lymph nodes, recognizing pathogens, while LDCs differentiate into typical DCs, mainly producing interferon (Ciaramella et al., 2013). Further studies suggest that peripheral DCs in PD mouse models may enter the brain, residing in the choroid plexus or meninges, engaging with various antigens to promote neuroinflammatory processes (Negi and Das 2018). Additionally, tolerogenic bone marrow-derived DCs induce neuroprotective Tregs in MPTP-induced PD mouse models (Schutt, Gendelman, and Mosley 2018). Remarkably, dendritic cell factor 1 (DCF1), a membrane protein crucial for nerve development in mice, can prevent α-synuclein-induced dopaminergic neuron loss by aggregating α-synuclein in the dorsomedial region of Drosophila (S. Zhang et al., 2018).

T cells can enter the CNS after experiencing damage, influencing the activation of glial cells and the degeneration of neurons (Mundt et al., 2019). APCs are recruited into the CSF in neurodegenerative diseases. These APCs present antigen peptides using MHC class II molecules, which then activate infiltrating T cells through the binding of their surface TCR to the presented antigen peptides. This interaction between TCR and MHC leads to the production of both CD4+ and CD8+ T cells (Z. Chen, Chen, and Liu 2018b). In individuals with PD, there is a decrease in T cell levels in peripheral blood (Álvarez-Luquín et al., 2019). Evidence from postmortem brains of PD patients supports T cell infiltration, which exacerbates neurotoxicity in a PD mouse model (Brochard et al., 2009). In a pivotal 2017 study by Sulzer et al., it was reported that in individuals with PD, particular T cell subsets, particularly CD4+ T cells, displayed recognition of specific α-synuclein peptides (Sulzer et al., 2017). Deletion of CD4+ lymphocytes in both MPTP and an α-synuclein model alleviated neurodegeneration (Williams et al., 2021). In a rat model of Parkinson's disease, the injection of α-synuclein led to T cell infiltration and microglial activation, contributing to dopaminergic neuron loss (Subbarayan et al., 2020). A more recent study has shown that the immunoreactivity of α-synuclein-specific T cells in peripheral blood mononuclear cells is linked to the preclinical and early stages of motor PD (Lindestam Arlehamn et al., 2020).

CD4+ FOXP3+ Treg frequency increased with age and was associated with enhanced suppressive activity in PD patients, suggesting their role in neuroinflammation (Rosenkranz et al., 2007). Cognitive impairment has been linked to a higher number of circulating lymphocytes and an imbalance in the compartment of Tregs (Z. X. Hu et al., 2018). Interestingly, elevated levels of IL-1β, TNF-α, IL-2, peripheral blood lymphocytes in the serum, and CSF of PD patients (Brochard et al., 2009). Tregs play a role in immunoregulation, bridging the peripheral and central immune systems. For instance, in an MPTP-induced PD mouse model, Tregs protected against the loss of dopaminergic neurons in the SN (K. H. Kim et al., 2019). Similarly, Tregs function appears to decrease in the periphery of a 6-OHDA-induced PD rat model, coinciding with T cell infiltration when neuronal loss in the SN reaches 80% (J.-J. Xiao et al. 2015).

Apart from this, in PD patients, there was an increase in the percentage of NK cells, but the activity of NK cells remained unchanged (Jiang et al., 2017). Multiple lines of evidence suggest a potential positive association between NK cell levels and the severity of PD (Green et al., 2017). Notably, NK cells were shown to have a modulatory effect on α-synuclein pathology and motor symptoms in an α-synuclein transgenic mouse model (Earls et al., 2020). Further, in individuals with PD, B cells are not typically found in the brain. However, deposits of IgG antibodies were discovered in dopaminergic neurons, and Lewy bodies were coated with IgG (Orr et al., 2005). Moreover, analysis of cell levels in the blood of PD patients showed that the reduction of CD4+ T helper cells and CD19+ B cells worsened with increasing clinical severity (Kobo et al., 2016; C. H. Stevens et al., 2012). An investigation of peripheral IgG+ memory B cells from PD patients, examining their reactivity to α-synuclein, revealed that naturally occurring antibodies derived from PD patients inhibited α-synuclein seeding in vitro and recognized Lewy pathology (X. Li et al., 2019b). Patients with PD exhibit the presence of autoantibodies against α-synuclein, dopamine, and melanin in both their serum and CSF (Double et al., 2009; Yanamandra et al., 2011). Correlations between α-synuclein autoantibody levels in the CSF and plasma of individuals with mild to moderate PD and disease activity support the potential use of α-synuclein autoantibodies as a biomarker for PD (Horvath et al., 2017).

PD patients exhibit faster progression and diminished responses to current treatments. The neutrophil-to-lymphocyte ratio (NLR) and platelet-to-lymphocyte ratio (PLR) are potential neuroinflammation biomarkers (Alster, Madetko, and Friedman 2021; Madetko et al., 2022). CSF is a valuable source for studying molecular changes in neurodegeneration. Inflammatory molecules in peripheral blood and CSF provide potential biomarkers (Qin et al., 2016). Blood biomarker leakage in PD patients suggests the involvement of the gut-brain axis. Possible biomarkers include IL-1β, IL-2, IL-6, IL-10, hsCRP, TNF-α, and sTNFRs (T.-W. Liu, Chen, and Chang 2022). Considering the extensive evidence of immune activation and inflammation in PD, it's not unexpected that clinical trials are exploring anti-inflammatory drugs and interventions aimed at the immune system to address the condition. Numerous anti-inflammatory treatments, including dexamethasone, ibuprofen, amantadine, and more, have demonstrated protective effects against dopaminergic cell death in animal models of PD (Manthripragada et al., 2011; X. Gao et al., 2011; Malagelada et al., 2010; Delgado, Jonakait, and Ganea 2002; Lamine et al., 2016). However, the effects of these drugs on PD patients are inconclusive. While one meta-analysis suggests that only ibuprofen may offer a modest protective effect (Samii et al., 2009), another indicates that nonsteroidal anti-inflammatory drugs (NSAIDs) (except aspirin) might have a protective impact (Gagne and Power 2010). Minocycline, despite showing promise in preclinical studies, failed to alter early PD progression in a clinical trial (Investigators 2006, 2008; Quintana, Pérez-Sánchez, and Farez 2014). In contrast, the glucagon-like peptide-1 receptor agonist NLY01 displayed neuroprotection in a sporadic PD mouse model by inhibiting the conversion of astrocytes into a neurotoxic phenotype mediated by microglia (Yun et al., 2018).

In the pursuit of therapeutic approaches for PD, active and passive immunization methods have been investigated. Active immunization stimulates the patient's immune system to produce antibodies against α-synuclein. Masliah et al., have shown promise in reducing toxic α-synuclein accumulation through active immunization in mouse models (Masliah et al., 2005). To minimize autoimmune risks, clinical trials have focused on using smaller α-synuclein peptide fragments as vaccines. Companies like Affiris have tested AFFITOPE® vaccines that mimic α-synuclein C-terminus, with positive results in preclinical and early clinical trials (Mandler et al., 2014; Volc et al., 2020). Passive immunization involves administering laboratory-generated antibodies directly targeting α-synuclein. Monoclonal antibodies have been developed to bind to various α-synuclein forms and have demonstrated potential in reducing α-synuclein accumulation and preserving neurons in animal models. Nanobodies, smaller antibody fragments, have also shown promise in preventing α-synuclein aggregation and reducing toxicity, with advantages like improved brain penetration and reduced immunogenicity (Castonguay, Gravel, and Lévesque 2021).

2.6 ROLE OF IMMUNE SYSTEM IN PolyQ DISORDERS

Polyglutamine-related expansion disorders is an umbrella term to encompass several heritable disorders including HD and spinocerebellar ataxias (SCA1, SCA2, SCA3, SCA6, SCA7, SCA17) (Cohen-Carmon and Meshorer 2012). When CAG repeats which code for glutamine tract (PolyQ) expands beyond a normal threshold, it becomes unstable, causing these disorders. In healthy individuals, the polyQ tract widely varies between 35–50 repeats, whereas, in carriers or patients, the polyQ tract reaches above 40 and is mostly higher than 100 glutamines (Cohen-Carmon and Meshorer 2012). The CAG repeat expansions take place during multiple phases during growth and development, however, an abnormal increase in the CAG repeats affects the neuronal function dynamics and exacerbates the pathophysiology of polyQ-related disorders (Cohen-Carmon and Meshorer 2012). While several hypotheses govern how this abnormal expansion leads to different classes of polyQ disorder, some of the most widely followed hypotheses are discussed below. First, the conformational change in the polyQ domain affects the intrinsic physiological property of the protein (Paulson, Bonini, and Roth 2000). For instance, the polyQ expansion in the SCA6 protein has been thought to affect the intrinsic physiological properties of the voltage-gated calcium channel. Second, the interaction between mutant protein and the associated interacting partners as well as spatio-temporal crosstalk in the CNS contributes to the selective vulnerability of specific neuronal subtypes (Paulson, Bonini, and Roth 2000). As a third possibility, mutant polyQ proteins may bind and sequester glutamine-rich transcription factors, altering neuronal gene expression critical for neuronal survival (Perez et al., 1998; Kazantsev et al., 1999).

A common integrating feature of polyQ-related disorders is the selective vulnerability of specific neuronal subtypes followed by reactive gliosis and neuroinflammation. Previous reports suggest a strong activation of immune response in polyQ disorders (Paulson, Bonini, and Roth 2000), HD being the most studied and reported (Olejniczak, Urbanek, and Krzyzosiak 2015). In HD, the CAG repeats expansion takes place in the exon 1 of the *HTT* gene which codes for the huntingtin protein. While in a healthy individual, the CAG repeats vary between 6 and 35 units, 36–39 repeats exhibit reduced penetrance, and more than 40 repeats are associated with HD development and pathophysiology (Duyao et al., 1993; Morrison et al., 2007; Group 1998). In HD, there is a selective vulnerability of neurons in the cortex and the striatum which leads to movement abnormalities, and dementia and turns fatal as the disease progresses (Olejniczak, Urbanek, and Krzyzosiak 2015). Previous clinical and post-mortem reports suggest a strong association of microglial activation in HD brain (Sapp et al., 2001; Tai et al., 2007; Crocker, Costain, and Robertson 2006). The level of microglial activation is strongly correlated with the progression of the disorder and strongly associated with the severity of the pathophysiology (Dalrymple et al., 2007; Pavese et al., 2006). The mutant HTT expressing microglia expresses increased levels of proinflammatory cytokines including IL-8, IL-6, and TNF-α which are often thought to impair the CNS homeostatic response to brain injury. The upregulation of the proinflammatory factors is caused by the proteasomal dysfunction induced by the mutant HTT (Björkqvist et al., 2008). Additionally, the mutant HTT also affects astrocytic regulation of neuronal synapses. In patients with HD, reactive astrocytes exhibit hypertrophy and upregulation of several molecules including GFAP, S100B, iNOS, and NFκB (J. Bradford et al., 2009; J. W. Bradford, Li, and Li 2010).

Moreover, astrocytes can be considered important targets in the progression of HD because HD models that do not express mutant HTT in neurons still have reactive gliosis as an important feature in their pathophysiology (J. Bradford et al., 2009). Similarly, another report suggests that the mutant huntingtin-mediated increase in IκB kinase (IKK) activity leads to an increase in astrocytic NFκB but not in microglia or neuronal cell types in the HD patients as well as rodent models of HD (Hsiao et al., 2013). HD rodent models have also shown caspases, nitric oxide, and iNOS as responsible factors mediating reactive astrogliosis and apoptosis (Ona et al., 1999; Tabrizi et al., 2000; M. Chen et al., 2000). Of particular interest is the caspase protein, which is often found to be regulatory of the pathological symptoms of HD. For instance, in the late presymptomatic phase of HD, caspase-1 activation has been found where, as the disease progresses, the caspase-3 levels are upregulated. It is interesting to note that huntingtin is a substrate for both caspase 1 and 3 (Wellington et al., 2002; Goldberg et al., 1996; Wellington et al., 1998; Mejia and Friedlander 2001). Additionally, the complement-mediated proteins, which are crucial in the complement pathway, have also been found to be upregulated in HD brains (Singhrao et al., 1999). Taken together, this suggests that at the CNS level, there is a close cross-talk between different neuronal and glial cell types, which affect the immune system activation in the HD pathophysiology.

The role of the peripheral immune system in HD pathophysiology has also been studied extensively (Olejniczak, Urbanek, and Krzyzosiak 2015). Primarily, with the progression of the HD pathology the level of mutant HTT in leukocytes has been found to increase and positively correlate as a disease progressor in these cells (Weiss et al., 2012). An elevated level of plasma cytokines in the HD patients, including TNF-α, IL-1β, IL-4, IL-10, IL-8, and IL-6, and chemokines MIP-1β, MCP-4, eotaxin-3, and MCP-1, have been observed (Dalrymple et al., 2007; Wild et al., 2011; Kwan et al., 2012). Previous reports strongly associate the immune cell activation as a primary effect of the mutant HTT expression instead of being a secondary response. Additionally, the CAG repeat length is significantly associated with the TNF-α levels produced by the HD monocytes (Träger et al., 2014). The activity of the NFκB transcription factor is modified by mutant huntingtin, and altered transcription of NFκB target genes resulted in enhanced cytokine (IL-6 and TNF-α) production (Hsiao et al., 2013).

As discussed earlier, several polyQ proteins act as substrates for the caspases. It is important to note that the truncated version of the protein is toxic and plays a crucial role in the pathogenesis of the respective polyQ disorder (Olejniczak, Urbanek, and Krzyzosiak 2015). The immune system activation in other polyQ disorders, such as the SCAs, is less explored (the majority of the work being in the SCA3 also known as the Machado-Joseph disease (MJD)) compared to the HDs (Olejniczak, Urbanek, and Krzyzosiak 2015). The MJD is caused by the expansion of CAG repeats from the physiological 10–51 repeats to pathological 55–87 repeats in the exon 10 of the *ATXN3* gene (Kawaguchi et al., 1994). The *ATXN3* gene encodes for the ataxin-3 enzyme which is involved in the proteasomal protein degradation pathway (Olejniczak, Urbanek, and Krzyzosiak 2015). Aggregates of mutant ataxin-3 are observed in the patients of MJD (Schmidt et al., 1998). Studies have identified differential expression of several immunoregulatory genes, including interleukin-1 receptor-related Fos-inducible transcript (*Fit-1S*), transmembrane protein amyloid precursor protein (*APP*), cytokine stromal cell-derived factor 1α (*SDF1α*) and endopeptidase matrix metalloproteinase 2 (*MMP-2*) in the rodent SCA3-Q70 cell line (Bernd O Evert et al., 2001, 2003; B O Evert et al., 2006). An associated increase in the protein levels of IL-1β, chemokine SDF1, and anti-inflammatory interleukin-1 receptor antagonist (IL-1ra) has also been shown in SCA3 pontine human neurons (Bernd O Evert et al., 2003). Additionally, an increased number of reactive astrocytes and activated microglia have been observed in SCA3 patients (Olejniczak, Urbanek, and Krzyzosiak 2015).

2.7 ROLE OF COMPLEMENT SYSTEM IN NEUROINFLAMMATION AND NEURODEGENERATION

The complement system is considered an essential component of the innate immunity defense cascade. It was initially discovered in 1891 and was reported as a heat-sensitive factor having the ability to destroy the bacteria in the blood (Buchner

1981; Ehrlich and Morgenroth 1899). Over the years in 1899, it was described to be a supplementary molecule necessary for mediating antibody-induced bacterial lysis. As time has progressed, our knowledge and understanding of the complement system has gone beyond its primary role as a supplementary molecule (Orsini et al., 2014). The complement system has been implicated in numerous neurological disorders, as well as neuronal injury. The complement system activation occurs via a response to exogenous or endogenous danger signals, which include either damage-associated molecular patterns (DAMPs) or pathogen-associated molecular patterns (PAMPs) (Orsini et al., 2014). This activation can occur via different types of danger signals through the classical (CP), leptin (LP), or the alternate pathway (AP). In the classical pathway, C1q binds directly to the antigen-antibody complex or is activated directly by β-amyloid, C reactive protein (CRP), and apoptosis (Orsini et al., 2014). The leptin pathway is mediated by several pattern recognition receptors, including ficolins, mannose-binding lectin (MBL), and collectin-11 (CL-11). On the other hand, the alternate pathway is mediated by the hydrolysis of circulating C3 (Orsini et al., 2014). The complement activation triggers the phagocytic response via the cleavage products such as C3b and C4b, which act as opsonins (Orsini et al., 2014). The complement system functions as an essential guard by interacting with immune cells. Complement proteins additionally have a scavenging role where they monitor the system for potential threats. Both neurons and glial subtypes can produce C1q and C3 complement initiator molecules, and receptors such as C3aR (Orsini et al., 2014). The complement-mediated response is highly associated with chronic and acute pathological scenarios and is often thought to drive the pathological state of certain neurodisorders (discussed later in this section).

The complement system plays a key role in maintaining synaptic homeostasis and synaptic remodeling. In this section, we will understand the role of the complement system in synaptic plasticity, development, and homeostasis. During the developmental process, neuronal circuits undergo several refinements and fine-tuning, which help in the maturation of the synapses. The primary synaptic refinement during the developmental phase has been widely studied using the retinogeniculate system. At early stages of development, retinal ganglion cells (RGCs) form transient synaptic connections to neurons of the dorsal lateral geniculate nucleus (dLGN) (Orsini et al., 2014). During the first phase of postnatal development, the synaptic circuits undergo rapid remodeling to ensure functional visual circuitry (Hooks and Chen 2006). Astrocytes play a crucial role in refining the synaptic circuitry and rendering mature functional synaptic connections. Additionally, astrocytes release TGF-β, which mediates the enhancement of C1q in the RGC, which is further transported to the dLGNs. C1q acts as "eat-me tags," which contributes either to direct synaptic tagging and elimination or, via C3b, opsonizes the synapses (Orsini et al., 2014). It is important to note that microglia express higher levels of C1q during the postnatal period, which coincides with the peak of synaptic remodeling and refinement (B. Stevens et al., 2007). Another important role of the complement system is to clear cellular debris and remove apoptotic cells (Ricklin et al., 2010). Previous studies have shown that C1q tags the neuronal apoptotic cells and activates the classical complement pathway leading to the production of C3b and C4b (Cole et al., 2006) responsible for the phagocytic removal of apoptotic cells (Dalakas, Alexopoulos, and Spaeth 2020; Schartz and Tenner 2020; Warwick et al., 2021).

C1q binds to endothelial surfaces and mediates the expression of several adhesion molecules and the release of chemokines and cytokines, including IL-8 and IL-6 (S. Xiao, Xu, and Jarvis 2001). Proteins involved in the complement pathways have been strongly implicated in a wide array of neurodisorders, including AD, PD as well as polyQ disorders such as HD (Reviewed in (Orsini et al., 2014)). Therapeutics targeting the complement system are now being widely explored for their therapeutic potential. Increased complement activation has been previously observed in Lewy bodies and oligodendroglia in SN in PD patients (Goldknopf et al., 2006; Patrick L McGeer and McGeer 2004). Components of the complement system, including C3c, C3d, C2, C3, and C1q, have been found in serum and spinal cord tissue in ALS patients (Grewal, Morgan, and Finch 1999; Tsuboi and Yamada 1994).

The role of the complement system is widely studied in AD pathophysiology. During synaptic maturation and synapse formation, complement proteins such as C1q and C3 normally associate with glial cells for the pruning of synapses. However, they are aberrantly activated in AD pathophysiology and drive neurotoxicity as the disease progresses (Pekna and Pekny 2021; Clarke and Barres 2013). Astrocytes express several phagocytic receptors, which include Multiple EGF Like Domains 10 (MEGF10) and Mer Tyrosine Kinase (MERTK) (Iram et al., 2016). The complement components such as C1q or C3 have been thought to drive the neurotoxicity and pathophysiological symptoms of AD. This is evident from the reports which show that C3, C1q, or microglial CR3 receptor inhibition rescues early synapse loss in AD (Hong et al., 2016). Additionally, the C1q knockout AD model has shown beneficial effects (Fonseca et al., 2004). In postmortem brains from AD patients, complement proteins C1q, C3, and C4 have been found to be colocalized with amyloid plaques (P L McGeer et al., 1989). The C1q is also thought to be an essential component that drives the neurotoxicity of soluble Aβ oligomers in impairing the hippocampal long-term potentiation (LTP) (Iram et al., 2016; Carpanini et al., 2022). Importantly, transgenic animals with reduced or compromised complement components show better protection against AD pathology (Shah, Kishore, and Shastri 2021). One such example is the study that shows a substantial reduction in activated microglia in transgenic AD mice (Tg2576 mice) crossed with homozygous C1q Knockout (KO) mice (Fonseca et al., 2004). The immunoreactivity of C1q is found to be elevated in rodent models of AD (Hong et al., 2016). Another study showed an increased association of C1q with postsynaptic markers when oligomeric Aβ is injected into WT mice, indicating an increased synaptic engulfment. All of these studies suggest that C1q in association with soluble Aβ in a positive feedforward loop influences the toxic pathophysiology of AD (Hong et al., 2016). Anti-C1q antibodies are now being explored as possible therapeutics. C1q deletion shows a marked improvement in AD pathology by reducing the association of astrocytes and synapses and thereby minimizing the loss of functional synapses (Dejanovic et al., 2022). Our recent work with the translocator protein ligand XBD173 shows that XBD173 provides procognitive benefits and rescues the synaptic density via a reduction in astrocytic C1q engulfment, both in the hippocampus and the cortex of the AD rodent model (Pradhan et al., 2023). In AD transgenic mice, we found a higher synaptic engulfment as well as higher C1q engulfment by the astrocytes, and a reduction of astrocytic C1q engulfment was associated with neuroprotective benefit by translocator protein ligand XBD173. Additionally, we reported for the first time a strong deposition of C1q in the form of aggregates at the amyloid plaque site in the transgenic Arc Aβ mice model of AD (Pradhan et al., 2023). Previously, amyloid β protein has been shown to be associated with complement component C3 within the substructural domains of drusen (Johnson et al., 2002). Another study has shown that the C4 binding protein and factor H of the complement pathway colocalize with amyloid plaques and dead cells in AD brains (Tenner 2020). While it is inconclusive to describe the components of the C1q aggregates that we found at the plaque site, a possible deposition of dystrophic neurites as well as sequestered complement components including C1q and C3 could not be ruled out. Another crucial finding from our study was that while all the C1q aggregates were associated with amyloid plaques, not all the plaques were associated with C1q aggregates (Pradhan et al., 2023). This possibly hints towards a more neurotoxic feature where the C1q aggregate colocalizes with the amyloid plaques. All these studies taken together highlight the importance of the complement system in mediating neuroinflammatory response and in mediating the pathophysiology of neurological disorders.

2.8 DISCUSSION

Neuroinflammation is a common pathological feature of nerve injury as well as neurological disorders. The immune system provides different lines of defense against existing pathological conditions. It is important to keep in mind that glial cells act as a surveillance system, continuously monitoring and refining the synapses. In a physiological scenario, astrocytes and microglia crosstalk with each other and refine the synaptic circuitry to modulate the formation of mature and functional synaptic connections. However, in a pathological state, the glial cells become reactive and activated. Let's understand this transition from a physiological to a pathological state from an energy standpoint. Maintenance of the energy ecosystem and homeostasis in the physiological state and their disturbance in the pathological state are most commonly overlooked by researchers. Understanding the energy balance provides us with an overall picture of the pathophysiological effects of neurodisorders. Synaptic pruning, the process by which adult neurons undergo a series of refining to form functional circuitry, is an energy-dependent process. Essentially, removing unwanted synapses betters signal transmission by removing irrelevant neurons, which saves energy. This process of removing unwanted connections however requires energy (Pradhan et al., 2024). Previous studies have highlighted that astrocytes in an ATP-dependent manner mediate the elimination of unwanted synapses via purinergic signaling pathway activation (Yang et al., 2016). This delicate interplay between energy balances governs the process of synaptic refinement during functional circuitry formation in a physiological state. However, this energy balance is heavily disturbed in neurodisorders such as AD. Reactive microglia and activated astrocytes in AD lead to an aberrant increase in functional synapse elimination, and this adds further to the energy crisis in AD. Maintaining energy homeostasis is therefore a critical factor in deciding the fate of neurons in neurodisorders.

REFERENCES

Abrahamsson, Sofia V, Daniela F Angelini, Amy N Dubinsky, Esther Morel, Unsong Oh, Joanne L Jones, Daniele Carassiti, Richard Reynolds, Marco Salvetti, and Peter A Calabresi. 2013. "Non-Myeloablative Autologous Haematopoietic Stem Cell Transplantation Expands Regulatory Cells and Depletes IL-17 Producing Mucosal-Associated Invariant T Cells in Multiple Sclerosis." *Brain* 136 (9): 2888–2903.

Ader, Robert, and Nicholas Cohen. 1975. "Behaviorally Conditioned Immunosuppression." *Psychosomatic Medicine* 37 (4): 333–40.

Aguzzi, Adriano, Ben A Barres, and Mariko L Bennett. 2013. "Microglia: Scapegoat, Saboteur, or Something Else?" *Science* 339 (6116): 156–61.

Alster, Piotr, Natalia Madetko, and Andrzej Friedman. 2021. "Neutrophil-to-Lymphocyte Ratio (NLR) at Boundaries of Progressive Supranuclear Palsy Syndrome (PSPS) and Corticobasal Syndrome (CBS)." *Neurologia i Neurochirurgia Polska* 55 (1): 97–101.

Álvarez-Luquín, Diana D, Asiel Arce-Sillas, Jaquelín Leyva-Hernández, Edgar Sevilla-Reyes, Marie Catherine Boll, Esteban Montes-Moratilla, Viridiana Vivas-Almazán, Citzielli Pérez-Correa, Ulises Rodríguez-Ortiz, and Raquel Espinoza-Cárdenas. 2019. "Regulatory Impairment in Untreated Parkinson's Disease Is Not Restricted to Tregs: Other Regulatory Populations are also Involved." *Journal of Neuroinflammation* 16: 1–11.

Annibali, Viviana, Giovanni Ristori, Daniela F Angelini, Barbara Serafini, Rosella Mechelli, Stefania Cannoni, Silvia Romano, Andrea Paolillo, Hadi Abderrahim, and Adamo Diamantini. 2011. "CD161highCD8+ T Cells Bear Pathogenetic Potential in Multiple Sclerosis." *Brain* 134 (2): 542–54.

Ascherio, Alberto, and Kassandra L Munger. 2007. "Environmental Risk Factors for Multiple Sclerosis. Part II: Noninfectious Factors." *Annals of Neurology: Official Journal of the American Neurological Association and the Child Neurology Society* 61 (6): 504–13.

Ascherio, Alberto, Kassandra L Munger, Evelyne T Lennette, Donna Spiegelman, Miguel A Hernán, Michael J Olek, Susan E Hankinson, and David J Hunter. 2001. "Epstein-Barr Virus Antibodies and Risk of Multiple Sclerosis: A Prospective Study." *Jama* 286 (24): 3083–88.

Bamberger, Maria E, Meera E Harris, Douglas R McDonald, Jens Husemann, and Gary E Landreth. 2003. "A Cell Surface Receptor Complex for Fibrillar β-Amyloid Mediates Microglial Activation." *Journal of Neuroscience* 23 (7): 2665–74.

Bar-Or, Amit. 2005. "Immunology of Multiple Sclerosis." *Neurologic Clinics* 23 (1): 149–75.

Bar-Or, Amit, Lama Fawaz, Boli Fan, Peter J Darlington, Aja Rieger, Christine Ghorayeb, Peter A Calabresi, Emmanuelle Waubant, Stephen L Hauser, and Jiameng Zhang. 2010. "Abnormal B-Cell Cytokine Responses a Trigger of T-Cell – Mediated Disease in MS?" *Annals of Neurology* 67 (4): 452–61.

Baranzini, Sergio E, Carita Elfstrom, Sheng-Yung Chang, Catalin Butunoi, Ronald Murray, Russell Higuchi, and Jorge R Oksenberg. 2000. "Transcriptional Analysis of Multiple Sclerosis Brain Lesions Reveals a Complex Pattern of Cytokine Expression." *The Journal of Immunology* 165 (11): 6576–82.

Bardehle, Sophia, Martin Krüger, Felix Buggenthin, Julia Schwausch, Jovica Ninkovic, Hans Clevers, Hugo J Snippert, Fabian J Theis, Melanie Meyer-Luehmann, and Ingo Bechmann. 2013. "Live Imaging of Astrocyte Responses to Acute Injury Reveals Selective Juxtavascular Proliferation." *Nature Neuroscience* 16 (5): 580–86.

Barkholt, P, V Sanchez-Guajardo, Deniz Kirik, and M Romero-Ramos. 2012. "Long-Term Polarization of Microglia Upon α-Synuclein Overexpression in Nonhuman Primates." *Neuroscience* 208: 85–96.

Bartels, A L, A T M Willemsen, J Doorduin, E F J De Vries, R A Dierckx, and K L Leenders. 2010. "[11C]-PK11195 PET: Quantification of Neuroinflammation and a Monitor of Anti-Inflammatory Treatment in Parkinson's Disease?" *Parkinsonism & Related Disorders* 16 (1): 57–59.

Biase, Lindsay M De, Kornel E Schuebel, Zachary H Fusfeld, Kamwing Jair, Isobel A Hawes, Raffaello Cimbro, Hai-Ying Zhang, Qing-Rong Liu, Hui Shen, and Zheng-Xiong Xi. 2017. "Local Cues Establish and Maintain Region-Specific Phenotypes of Basal Ganglia Microglia." *Neuron* 95 (2): 341–56.

Biddison, William E, William W Cruikshank, David M Center, Clara M Pelfrey, Dennis D Taub, and Richard V Turner. 1998. "CD8+ Myelin Peptide-Specific T Cells can Chemoattract CD4+ Myelin Peptide-Specific T Cells: Importance of IFN-Inducible Protein 10." *The Journal of Immunology* 160 (1): 444–48.

Bielekova, Bibiana, Marta Catalfamo, Susan Reichert-Scrivner, Amy Packer, Magdalena Cerna, Thomas A Waldmann, Henry McFarland, Pierre A Henkart, and Roland Martin. 2006. "Regulatory CD56bright Natural Killer Cells Mediate Immunomodulatory Effects of IL-2Rα-Targeted Therapy (Daclizumab) in Multiple Sclerosis." *Proceedings of the National Academy of Sciences* 103 (15): 5941–46.

Björkqvist, Maria, Edward J Wild, Jenny Thiele, Aurelio Silvestroni, Ralph Andre, Nayana Lahiri, Elsa Raibon, Richard V Lee, Caroline L Benn, and Denis Soulet. 2008. "A Novel Pathogenic Pathway of Immune Activation Detectable Before Clinical Onset in Huntington's Disease." *The Journal of Experimental Medicine* 205 (8): 1869–77.

Bliederhaeuser, Corinna, Lisa Zondler, Veselin Grozdanov, Wolfgang P Ruf, David Brenner, Heather L Melrose, Peter Bauer, Albert C Ludolph, Frank Gillardon, and Jan Kassubek. 2016. "LRRK2 Contributes to Monocyte Dysregulation in Parkinson's Disease." *Acta Neuropathologica Communications* 4: 1–5.

Booth, Frank W, Christian K Roberts, John P Thyfault, Gregory N Ruegsegger, and Ryan G Toedebusch. 2017a. "Role of Inactivity in Chronic Diseases: Evolutionary Insight and Pathophysiological Mechanisms." *Physiological Reviews* 97 (4): 1351–402.

Booth, Heather D E, Warren D Hirst, and Richard Wade-Martins. 2017b. "The Role of Astrocyte Dysfunction in Parkinson's Disease Pathogenesis." *Trends in Neurosciences* 40 (6): 358–70.

Bourque, Jessica, and Daniel Hawiger. 2021. "Current and Future Immunotherapies for Multiple Sclerosis." *Missouri Medicine* 118 (4): 334.

Bradford, Jennifer, Ji-Yeon Shin, Meredith Roberts, Chuan-En Wang, Xiao-Jiang Li, and Shihua Li. 2009. "Expression of Mutant Huntingtin in Mouse Brain Astrocytes Causes Age-Dependent Neurological Symptoms." *Proceedings of the National Academy of Sciences* 106 (52): 22480–85.

Bradford, Jennifer W, Shihua Li, and Xiao-Jiang Li. 2010. "Polyglutamine Toxicity in Non-Neuronal Cells." *Cell Research* 20 (4): 400–07.

Britschgi, M, C E Olin, H T Johns, Y Takeda-Uchimura, M C LeMieux, K Rufibach, J Rajadas, H Zhang, B Tomooka, and W H Robinson. 2009. "Neuroprotective Natural Antibodies to Assemblies of Amyloidogenic Peptides Decrease with Normal Aging and Advancing Alzheimer's Disease." *Proceedings of the National Academy of Sciences* 106 (29): 12145–50.

Brochard, Vanessa, Béhazine Combadière, Annick Prigent, Yasmina Laouar, Aline Perrin, Virginie Beray-Berthat, Olivia Bonduelle, Daniel Alvarez-Fischer, Jacques Callebert, and Jean-Marie Launay. 2009. "Infiltration of CD4+ Lymphocytes into the Brain Contributes to Neurodegeneration in a Mouse Model of Parkinson Disease." *The Journal of Clinical Investigation* 119 (1):182–92.

Buchner, H. 1981. "Centr Bakteriol Parasitenk." *Zur. Nomenkl. Schutzenden Eiweisskorper* 10: 699–701.

Cabezas, Ricardo, Marcos Ávila, Janneth Gonzalez, Ramon Santos El-Bachá, Eliana Báez, Luis Miguel García-Segura, Juan Camilo Jurado Coronel, Francisco Capani, Gloria Patricia Cardona-Gomez, and George E Barreto. 2014. "Astrocytic Modulation of Blood Brain Barrier: Perspectives on Parkinson's Disease." *Frontiers in Cellular Neuroscience* 8: 211.

Cao, Shuwen, David G Standaert, and Ashley S Harms. 2012. "The Gamma Chain Subunit of Fc Receptors Is Required for Alpha-Synuclein-Induced Pro-Inflammatory Signaling in Microglia." *Journal of Neuroinflammation* 9: 1–11.

Carpanini, Sarah M, Megan Torvell, Ryan J Bevan, Robert A J Byrne, Nikoleta Daskoulidou, Takashi Saito, Takaomi C Saido, Philip R Taylor, Timothy R Hughes, and Wioleta M Zelek. 2022. "Terminal Complement Pathway Activation Drives Synaptic Loss in Alzheimer's Disease Models." *Acta Neuropathologica Communications* 10 (1): 1–16.

Carton, Herwig, Robert Vlietinck, Jan Debruyne, Jacques De Keyser, Marie-Beatrice D'Hooghe, Ruth Loos, R Medaer, L Truyen, I M Yee, and A D Sadovnick. 1997. "Risks of Multiple Sclerosis in Relatives of Patients in Flanders, Belgium." *Journal of Neurology, Neurosurgery & Psychiatry* 62 (4): 329–33.

Castonguay, Anne-Marie, Claude Gravel, and Martin Lévesque. 2021. "Treating Parkinson's Disease with Antibodies: Previous Studies and Future Directions." *Journal of Parkinson's Disease* 11 (1): 71–92.

Castro-Sánchez, Sara, Ángel J García-Yagüe, Tresa López-Royo, Maria Casarejos, José Luis Lanciego, and Isabel Lastres-Becker. 2018. "Cx3cr1-Deficiency Exacerbates Alpha-Synuclein-A53T Induced Neuroinflammation and Neurodegeneration in a Mouse Model of Parkinson's Disease." *Glia* 66 (8): 1752–62.

Chakrabarty, Paramita, Carolina Ceballos-Diaz, Amanda Beccard, Christopher Janus, Dennis Dickson, Todd E Golde, and Pritam Das. 2010. "IFN-γ Promotes Complement Expression and Attenuates Amyloid Plaque Deposition in Amyloid β Precursor Protein Transgenic Mice." *The Journal of Immunology* 184 (9): 5333–43.

Chen, Cang, Xiuhua Li, Guo Ge, Jingwei Liu, K C Biju, Suzette D Laing, Yusheng Qian, Cori Ballard, Zhixu He, and Eliezer Masliah. 2018a. "GDNF-Expressing Macrophages Mitigate Loss of Dopamine Neurons and Improve Parkinsonian Symptoms in MitoPark Mice." *Scientific Reports* 8 (1): 5460.

Chen, Minghua, Victor O Ona, Mingwei Li, Robert J Ferrante, Klaus B Fink, Shan Zhu, Jie Bian, Lei Guo, Laurie A Farrell, and Steve M Hersch. 2000. "Minocycline Inhibits Caspase-1 and Caspase-3 Expression and Delays Mortality in a Transgenic Mouse Model of Huntington Disease." *Nature Medicine* 6 (7): 797–801.

Chen, Zhichun, Shengdi Chen, and Jun Liu. 2018b. "The Role of T Cells in the Pathogenesis of Parkinson's Disease." *Progress in Neurobiology* 169: 1–23.

Chiu, Isaac M, Emiko T A Morimoto, Hani Goodarzi, Jennifer T Liao, Sean O'Keeffe, Hemali P Phatnani, Michael Muratet, Michael C Carroll, Shawn Levy, and Saeed Tavazoie. 2013. "A Neurodegeneration-Specific Gene-Expression Signature of Acutely Isolated Microglia from an Amyotrophic Lateral Sclerosis Mouse Model." *Cell Reports* 4 (2): 385–401.

Cho, Seo-Hyun, Binggui Sun, Yungui Zhou, Tiina M Kauppinen, Brian Halabisky, Paul Wes, Richard M Ransohoff, and Li Gan. 2011. "CX3CR1 Protein Signaling Modulates Microglial Activation and Protects against Plaque-Independent Cognitive Deficits in a Mouse Model of Alzheimer Disease." *Journal of Biological Chemistry* 286 (37): 32713–22.

Choi, Dong-Joo, Hajie Yang, Sushil Gaire, Keon Ah Lee, Jiawei An, Byung Gon Kim, Ilo Jou, Sang Myun Park, and Eun-Hye Joe. 2020. "Critical Roles of Astrocytic-CCL2-dependent Monocyte Infiltration in a DJ-1 Knockout Mouse Model of Delayed Brain Repair." *Glia* 68 (10): 2086–101.

Choi, Sang-Ho, Saba Aid, Hyung-Wook Kim, Sharon H Jackson, and Francesca Bosetti. 2012. "Inhibition of NADPH Oxidase Promotes Alternative and Anti-inflammatory Microglial Activation during Neuroinflammation." *Journal of Neurochemistry* 120 (2): 292–301.

Ciaramella, Antonio, Francesca Salani, Federica Bizzoni, Francesco E Pontieri, Alessandro Stefani, Mariangela Pierantozzi, Francesca Assogna, Carlo Caltagirone, Gianfranco Spalletta, and Paola Bossu. 2013. "Blood Dendritic Cell Frequency Declines in Idiopathic Parkinson's Disease and Is Associated with Motor Symptom Severity." *PLoS One* 8 (6): e65352.

Clarke, Laura E, and Ben A Barres. 2013. "Emerging Roles of Astrocytes in Neural Circuit Development." *Nature Reviews Neuroscience* 14 (5): 311–21.

Cohen-Carmon, Dorit, and Eran Meshorer. 2012. "Polyglutamine (PolyQ) Disorders: The Chromatin Connection." *Nucleus* 3 (5): 433–41.

Cole, Duncan S, Timothy R Hughes, Philippe Gasque, and B Paul Morgan. 2006. "Complement Regulator Loss on Apoptotic Neuronal Cells Causes Increased Complement Activation and Promotes Both Phagocytosis and Cell Lysis." *Molecular Immunology* 43 (12): 1953–64.

Cook, D A, G T Kannarkat, A F Cintron, Laura M Butkovich, Kyle B Fraser, J Chang, N Grigoryan, S A Factor, Andrew B West, and J M Boss. 2017. "LRRK2 Levels in Immune Cells Are Increased in Parkinson's Disease." *NPJ Parkinson's Disease* 3 (1): 11.

Cotrina, Maria Luisa, and Maiken Nedergaard. 2002. "Astrocytes in the Aging Brain." *Journal of Neuroscience Research* 67 (1): 1–10.

Crocker, Susan F, Willard J Costain, and Harold A Robertson. 2006. "DNA Microarray Analysis of Striatal Gene Expression in Symptomatic Transgenic Huntington's Mice (R6/2) Reveals Neuroinflammation and Insulin Associations." *Brain Research* 1088 (1): 176–86.

Cunningham, Christopher L, Verónica Martínez-Cerdeño, and Stephen C Noctor. 2013. "Microglia Regulate the Number of Neural Precursor Cells in the Developing Cerebral Cortex." *Journal of Neuroscience* 33 (10): 4216–33.

Dalakas, Marinos C, Harry Alexopoulos, and Peter J Spaeth. 2020. "Complement in Neurological Disorders and Emerging Complement-Targeted Therapeutics." *Nature Reviews Neurology* 16 (11): 601–17.

Dalrymple, Annette, Edward J Wild, Richard Joubert, Kirupa Sathasivam, Maria Björkqvist, Åsa Petersén, Graham S Jackson, Jeremy D Isaacs, Mark Kristiansen, and Gillian P Bates. 2007. "Proteomic Profiling of Plasma in Huntington's Disease Reveals Neuroinflammatory Activation and Biomarker Candidates." *Journal of Proteome Research* 6 (7): 2833–40.

Dauer, William, and Serge Przedborski. 2003. "Parkinson's Disease: Mechanisms and Models." *Neuron* 39 (6): 889–909.

Davalos, Dimitrios, Jaime Grutzendler, Guang Yang, Jiyun V Kim, Yi Zuo, Steffen Jung, Dan R Littman, Michael L Dustin, and Wen-Biao Gan. 2005. "ATP Mediates Rapid Microglial Response to Local Brain Injury in Vivo." *Nature Neuroscience* 8 (6): 752–58.

Dejanovic, Borislav, Tiffany Wu, Ming-Chi Tsai, David Graykowski, Vineela D Gandham, Christopher M Rose, Corey E Bakalarski, Hai Ngu, Yuanyuan Wang, and Shristi Pandey. 2022. "Complement C1q-Dependent Excitatory and Inhibitory Synapse Elimination by Astrocytes and Microglia in Alzheimer's Disease Mouse Models." *Nature Aging* 2 (9): 837–50.

Delgado, Mario, G Miller Jonakait, and Doina Ganea. 2002. "Vasoactive Intestinal Peptide and Pituitary Adenylate Cyclase-activating Polypeptide Inhibit Chemokine Production in Activated Microglia." *Glia* 39 (2): 148–61.

Dendrou, Calliope A, Lars Fugger, and Manuel A Friese. 2015. "Immunopathology of Multiple Sclerosis." *Nature Reviews Immunology* 15 (9): 545–58.

Dickson, Dennis W. 2012. "Parkinson's Disease and Parkinsonism: Neuropathology." *Cold Spring Harbor Perspectives in Medicine* 2 (8).

Didonna, Alessandro, and Jorge R Oksenberg. 2017. "The Genetics of Multiple Sclerosis." In, edited by Ian S Zagon and Patricia J McLaughlin. Brisbane (AU). https://doi.org/10.15586/codon.multiplesclerosis.2017.ch1

Dilokthornsakul, Piyameth, Robert J Valuck, Kavita V Nair, John R Corboy, Richard R Allen, and Jonathan D Campbell. 2016. "Multiple Sclerosis Prevalence in the United States Commercially Insured Population." *Neurology* 86 (11): 1014–21.

Dimou, Leda, and Magdalena Götz. 2014. "Glial Cells as Progenitors and Stem Cells: New Roles in the Healthy and Diseased Brain." *Physiological Reviews* 94 (3): 709–37.

Dimou, Leda, Christiane Simon, Frank Kirchhoff, Hirohide Takebayashi, and Magdalena Götz. 2008. "Progeny of Olig2-Expressing Progenitors in the Gray and White Matter of the Adult Mouse Cerebral Cortex." *Journal of Neuroscience* 28 (41): 10434–42.

Dodel, Richard, Karthikeyan Balakrishnan, Kathy Keyvani, Oliver Deuster, Frauke Neff, Luminita-Cornelia Andrei-Selmer, Stephan Röskam, Carsten Stüer, Yousef Al-Abed, and Carmen Noelker. 2011. "Naturally Occurring Autoantibodies against β-Amyloid: Investigating Their Role in Transgenic Animal and in Vitro Models of Alzheimer's Disease." *Journal of Neuroscience* 31 (15): 5847–54.

Domenico, Angelique di, Giulia Carola, Carles Calatayud, Meritxell Pons-Espinal, Juan Pablo Muñoz, Yvonne Richaud-Patin, Irene Fernandez-Carasa, Marta Gut, Armida Faella, and Janani Parameswaran. 2019. "Patient-Specific IPSC-Derived Astrocytes Contribute to Non-Cell-Autonomous Neurodegeneration in Parkinson's Disease." *Stem Cell Reports* 12 (2): 213–29.

Dorsey, E, Todd Sherer, Michael S Okun, and Bastiaan R Bloem. 2018. "The Emerging Evidence of the Parkinson Pandemic." *Journal of Parkinson's Disease* 8 (s1): S3–8.

Double, Kay L, Dominic B Rowe, Francine M Carew-Jones, M Hayes, Daniel Kam Yin Chan, J Blackie, A Corbett, Ronald Joffe, V S Fung, and J Morris. 2009. "Anti-Melanin Antibodies Are Increased in Sera in Parkinson's Disease." *Experimental Neurology* 217 (2): 297–301.

Duyao, M, C Ambrose, R Myers, A Novelletto, F Persichetti, M Frontali, S Folstein, C Ross, M Franz, and M Abbott. 1993. "Trinucleotide Repeat Length Instability and Age of Onset in Huntington's Disease." *Nature Genetics* 4 (4): 387–92.

Dzamko, Nicolas, Amanda M Gysbers, Rina Bandopadhyay, Marc F Bolliger, Akiko Uchino, Ye Zhao, Masaki Takao, Sandrine Wauters, Wilma D J van de Berg, and Junko Takahashi-Fujigasaki. 2017. "LRRK2 Levels and Phosphorylation in Parkinson's Disease Brain and Cases with Restricted Lewy Bodies." *Movement Disorders* 32 (3): 423–32.

Earls, Rachael H, Kelly B Menees, Jaegwon Chung, Claire-Anne Gutekunst, Hyun Joon Lee, Manuel G Hazim, Balázs Rada, Levi B Wood, and Jae-Kyung Lee. 2020. "NK Cells Clear α-Synuclein and the Depletion of NK Cells Exacerbates Synuclein Pathology in a Mouse Model of α-Synucleinopathy." *Proceedings of the National Academy of Sciences* 117 (3): 1762–71.

Ehrlich, Paul, and Julius Morgenroth. 1899. "Zur Theorie Der Lysinwirkung." *Berliner Klinische Wochenschrift* 1: 6–9.

Engelhardt, Britta. 2008. "The Blood-Central Nervous System Barriers Actively Control Immune Cell Entry into the Central Nervous System." *Current Pharmaceutical Design* 14 (16): 1555–65.

Evert, B O, J Schelhaas, H Fleischer, R A De Vos, E R Brunt, W Stenzel, T Klockgether, and U Wüllner. 2006. "Neuronal Intranuclear Inclusions, Dysregulation of Cytokine Expression and Cell Death in Spinocerebellar Ataxia Type 3." *Clinical Neuropathology* 25 (6): 272–81.

Evert, Bernd O, Ina R Vogt, Claudia Kindermann, Lucia Ozimek, Rob A I de Vos, Ewout R P Brunt, Ina Schmitt, Thomas Klockgether, and Ullrich Wüllner. 2001. "Inflammatory Genes Are Upregulated in Expanded Ataxin-3-Expressing Cell Lines and Spinocerebellar Ataxia Type 3 Brains." *Journal of Neuroscience* 21 (15): 5389–96.

Evert, Bernd O, Ina R Vogt, Ana M Vieira-Saecker, Lucia Ozimek, Rob A I de Vos, Ewout R P Brunt, Thomas Klockgether, and Ullrich Wüllner. 2003. "Gene Expression Profiling in Ataxin-3 Expressing Cell Lines Reveals Distinct Effects of Normal and Mutant Ataxin-3." *Journal of Neuropathology & Experimental Neurology* 62 (10): 1006–18.

Ferretti, M T, M Merlini, C Späni, C Gericke, N Schweizer, Gaby Enzmann, Britta Engelhardt, L Kulic, T Suter, and R M Nitsch. 2016. "T-Cell Brain Infiltration and Immature Antigen-Presenting Cells in Transgenic Models of Alzheimer's Disease like Cerebral Amyloidosis." *Brain, Behavior, and Immunity* 54: 211–25.

Fogdell-Hahn, A, A Ligers, M Grønning, J Hillert, and O Olerup. 2000. "Multiple Sclerosis: A Modifying Influence of HLA Class I Genes in an HLA Class II Associated Autoimmune Disease." *Tissue Antigens* 55 (2): 140–48.

Fonseca, Maria Isabel, Jun Zhou, Marina Botto, and Andrea J Tenner. 2004. "Absence of C1q Leads to Less Neuropathology in Transgenic Mouse Models of Alzheimer's Disease." *Journal of Neuroscience* 24 (29): 6457–65.

Fourgeaud, Lawrence, Paqui G Través, Yusuf Tufail, Humberto Leal-Bailey, Erin D Lew, Patrick G Burrola, Perri Callaway, Anna Zagórska, Carla V Rothlin, and Axel Nimmerjahn. 2016. "TAM Receptors Regulate Multiple Features of Microglial Physiology." *Nature* 532 (7598): 240–44.

Frischer, Josa M, Stephan Bramow, Assunta Dal-Bianco, Claudia F Lucchinetti, Helmut Rauschka, Manfred Schmidbauer, Henning Laursen, Per Soelberg Sorensen, and Hans Lassmann. 2009. "The Relation between Inflammation and Neurodegeneration in Multiple Sclerosis Brains." *Brain* 132 (5): 1175–89.

Frost, Jeffrey L, and Dorothy P Schafer. 2016. "Microglia: Architects of the Developing Nervous System." *Trends in Cell Biology* 26 (8): 587–97.

Fuhrmann, Martin, Tobias Bittner, Christian K E Jung, Steffen Burgold, Richard M Page, Gerda Mitteregger, Christian Haass, Frank M LaFerla, Hans Kretzschmar, and Jochen Herms. 2010. "Microglial Cx3cr1 Knockout Prevents Neuron Loss in a Mouse Model of Alzheimer's Disease." *Nature Neuroscience* 13 (4): 411–13.

Furman, Jennifer L, Diana M Sama, John C Gant, Tina L Beckett, M Paul Murphy, Adam D Bachstetter, Linda J Van Eldik, and Christopher M Norris. 2012. "Targeting Astrocytes Ameliorates Neurologic Changes in a Mouse Model of Alzheimer's Disease." *Journal of Neuroscience* 32 (46): 16129–40.

Gagne, Joshua J, and Melinda C Power. 2010. "Anti-Inflammatory Drugs and Risk of Parkinson Disease: A Meta-Analysis." *Neurology* 74 (12): 995–1002.

Galea, Ian, Ingo Bechmann, and V Hugh Perry. 2007. "What Is Immune Privilege (Not)?" *Trends in Immunology* 28 (1): 12–18.

Gan, Li, Mark R Cookson, Leonard Petrucelli, and Albert R La Spada. 2018. "Converging Pathways in Neurodegeneration, from Genetics to Mechanisms." *Nature Neuroscience* 21 (10): 1300–1309.

Gao, Liang, David Brenner, Enric Llorens-Bobadilla, Gonzalo Saiz-Castro, Tobias Frank, Peter Wieghofer, Oliver Hill, Meinolf Thiemann, Saoussen Karray, and Marco Prinz. 2015. "Infiltration of Circulating Myeloid Cells through CD95L Contributes to Neurodegeneration in Mice." *Journal of Experimental Medicine* 212 (4): 469–80.

Gao, Xiang, Honglei Chen, Michael A Schwarzschild, and Alberto Ascherio. 2011. "Use of Ibuprofen and Risk of Parkinson Disease." *Neurology* 76 (10): 863–69.

Gaskin, F, J Finley, Q.I.A.N.G. Fang, S.H.I.H.U.A. Xu, and S.M. Fu. 1993. "Human Antibodies Reactive with Beta-Amyloid Protein in Alzheimer's Disease." *The Journal of Experimental Medicine* 177 (4): 1181–86.

Gee, Jillian R, and Jeffrey N Keller. 2005. "Astrocytes: Regulation of Brain Homeostasis via Apolipoprotein E." *The International Journal of Biochemistry & Cell Biology* 37 (6): 1145–50.

"Genetic Risk and a Primary Role for Cell-Mediated Immune Mechanisms in Multiple Sclerosis." 2011. *Nature* 476 (7359): 214–19.

Gericke, Christoph, Anna Mallone, Britta Engelhardt, Roger M Nitsch, and Maria Teresa Ferretti. 2020. "Oligomeric Forms of Human Amyloid-Beta (1–42) Inhibit Antigen Presentation." *Frontiers in Immunology* 11: 1029.

Ghosh, Simantini, Michael D Wu, Solomon S Shaftel, Stephanos Kyrkanides, Frank M LaFerla, John A Olschowka, and M Kerry O'Banion. 2013. "Sustained Interleukin 1β Overexpression Exacerbates Tau Pathology despite Reduced Amyloid Burden in an Alzheimer's Mouse Model." *Journal of Neuroscience* 33 (11): 5053–64.

Giuliani, Fabrizio, Cynthia G Goodyer, Jack P Antel, and V Wee Yong. 2003. "Vulnerability of Human Neurons to T Cell-Mediated Cytotoxicity." *The Journal of Immunology* 171 (1): 368–79.

Goldberg, Y Pl, Daw Nicholson, D M Rasper, M A Kalchman, H B Koide, R K Graham, M Bromm, P Kazemi-Esfarjani, N A Thornberry, and J P Vaillancourt. 1996. "Cleavage of Huntingtin by Apopain, a Proapoptotic Cysteine Protease, Is Modulated by the Polyglutamine Tract." *Nature Genetics* 13 (4): 442–49.

Goldenberg, M M. 2012. "Multiple Sclerosis Review." *Pharmacy and Therapeutics* 37 (3): 175–84."

Goldknopf, Ira L, Essam A Sheta, Jennifer Bryson, Brian Folsom, Chris Wilson, Jeff Duty, Albert A Yen, and Stanley H Appel. 2006. "Complement C3c and Related Protein Biomarkers in Amyotrophic Lateral Sclerosis and Parkinson's Disease." *Biochemical and Biophysical Research Communications* 342 (4): 1034–39.

Goldmann, Tobias, Peter Wieghofer, Marta Joana Costa Jordão, Fabiola Prutek, Nora Hagemeyer, Kathrin Frenzel, Lukas Amann, Ori Staszewski, Katrin Kierdorf, and Martin Krueger. 2016. "Origin, Fate and Dynamics of Macrophages at Central Nervous System Interfaces." *Nature Immunology* 17 (7): 797–805.

Gourraud, Pierre-Antoine, Hanne F Harbo, Stephen L Hauser, and Sergio E Baranzini. 2012. "The Genetics of Multiple Sclerosis: An Up-to-date Review." *Immunological Reviews* 248 (1): 87–103.

Graeber, Manuel B. 2010. "Changing Face of Microglia." *Science* 330 (6005): 783–88.

Grassivaro, Francesca, Ramesh Menon, Massimo Acquaviva, Linda Ottoboni, Francesca Ruffini, Andrea Bergamaschi, Luca Muzio, Cinthia Farina, and Gianvito Martino. 2020. "Convergence between Microglia and Peripheral Macrophages Phenotype during Development and Neuroinflammation." *Journal of Neuroscience* 40 (4): 784–95.

Gray, Madison T, and John M Woulfe. 2015. "Striatal Blood–Brain Barrier Permeability in Parkinson's Disease." *Journal of Cerebral Blood Flow & Metabolism* 35 (5): 747–50.

Green, Holly, Xiaoqun Zhang, Katarina Tiklova, Nikolaos Volakakis, Lovisa Brodin, Louise Berg, Paul Greengard, Thomas Perlmann, and Per Svenningsson. 2017. "Alterations of P11 in Brain Tissue and Peripheral Blood Leukocytes in Parkinson's Disease." *Proceedings of the National Academy of Sciences* 114 (10): 2735–40.

Grewal, Raji P, Todd E Morgan, and Caleb E Finch. 1999. "C1qB and Clusterin MRNA Increase in Association with Neurodegeneration in Sporadic Amyotrophic Lateral Sclerosis." *Neuroscience Letters* 271 (1): 65–7.

Grigoriadis, Nikolaos, and Vincent Van Pesch. 2015. "A Basic Overview of Multiple Sclerosis Immunopathology." *European Journal of Neurology* 22: 3–13.

Group, American College of Medical Genetics/American Society of Human Genetics Huntington Disease Genetic Testing Working. 1998. "ACMG/ASHG Statement. Laboratory Guidelines for Huntington Disease Genetic Testing." *American Journal of Human Genetics* 62 (5): 1243–7.

Grozdanov, Veselin, Corinna Bliederhaeuser, Wolfgang P Ruf, Valerie Roth, Kathrin Fundel-Clemens, Lisa Zondler, David Brenner, Ana Martin-Villalba, Bastian Hengerer, and Jan Kassubek. 2014. "Inflammatory Dysregulation of Blood Monocytes in Parkinson's Disease Patients." *Acta Neuropathologica* 128: 651–63.

Gu, Ben J, Xin Huang, Amber Ou, Alan Rembach, Christopher Fowler, Pavan K Avula, Adam Horton, James D Doecke, Victor L Villemagne, and S Lance Macaulay. 2016. "Innate Phagocytosis by Peripheral Blood Monocytes Is Altered in Alzheimer's Disease." *Acta Neuropathologica* 132: 377–89.

Guo, Shenrui, Hui Wang, and Yafu Yin. 2022. "Microglia Polarization from M1 to M2 in Neurodegenerative Diseases." *Frontiers in Aging Neuroscience* 14: 815347.

Hagemeyer, Nora, Klara-Maria Hanft, Maria-Anna Akriditou, Nicole Unger, Eun S Park, E Richard Stanley, Ori Staszewski, Leda Dimou, and Marco Prinz. 2017. "Microglia Contribute to Normal Myelinogenesis and to Oligodendrocyte Progenitor Maintenance during Adulthood." *Acta Neuropathologica* 134: 441–58.

Han, Xiaojuan, Sifan Sun, Yiming Sun, Qiqi Song, Jialei Zhu, Nanshan Song, Miaomiao Chen, Ting Sun, Meiling Xia, and Jianhua Ding. 2019. "Small Molecule-Driven NLRP3 Inflammation Inhibition via Interplay between Ubiquitination and Autophagy: Implications for Parkinson Disease." *Autophagy* 15 (11): 1860–81.

Hansen, David V, Jesse E Hanson, and Morgan Sheng. 2018. "Microglia in Alzheimer's Disease." *Journal of Cell Biology* 217 (2): 459–72.

Hauser, Stephen L. 1994. "Multiple Sclerosis and Other Demyelinating Diseases." *Harrison's Principles of Internal Medicine* 2287-2295.

Haynes, Sharon E, Gunther Hollopeter, Guang Yang, Dana Kurpius, Michael E Dailey, Wen-Biao Gan, and David Julius. 2006. "The P2Y12 Receptor Regulates Microglial Activation by Extracellular Nucleotides." *Nature Neuroscience* 9 (12): 1512–19.

Held, Kathrin, Eduardo Beltran, Markus Moser, Reinhard Hohlfeld, and Klaus Dornmair. 2015. "T-Cell Receptor Repertoire of Human Peripheral CD161hiTRAV1-2+ MAIT Cells Revealed by next Generation Sequencing and Single Cell Analysis." *Human Immunology* 76 (9): 607–14.

Heneka, Michael T, Markus P Kummer, Andrea Stutz, Andrea Delekate, Stephanie Schwartz, Ana Vieira-Saecker, Angelika Griep, Daisy Axt, Anita Remus, and Te-Chen Tzeng. 2013. "NLRP3 Is Activated in Alzheimer's Disease and Contributes to Pathology in APP/PS1 Mice." *Nature* 493 (7434): 674–78.

Herbst, Susanne, Philip Campbell, John Harvey, Elliott M Bernard, Venizelos Papayannopoulos, Nicholas W Wood, Huw R Morris, and Maximiliano G Gutierrez. 2020. "LRRK 2 Activation Controls the Repair of Damaged Endomembranes in Macrophages." *The EMBO Journal* 39 (18): e104494.

Hestvik, Anne Lise K. 2010. "The Double-Edged Sword of Autoimmunity: Lessons from Multiple Sclerosis." *Toxins* 2 (4): 856–77.

Hollenbach, Jill A, Marcelo J Pando, Stacy J Caillier, Pierre-Antoine Gourraud, and Jorge R Oksenberg. 2016. "The Killer Immunoglobulin-like Receptor KIR3DL1 in Combination with HLA-Bw4 Is Protective against Multiple Sclerosis in African Americans." *Genes & Immunity* 17 (3): 199–202.

Holmøy, Trygve, and Anne Lise K Hestvik. 2008. "Multiple Sclerosis: Immunopathogenesis and Controversies in Defining the Cause." *Current Opinion in Infectious Diseases* 21 (3): 271–78.

Hong, Soyon, Victoria F Beja-Glasser, Bianca M Nfonoyim, Arnaud Frouin, Shaomin Li, Saranya Ramakrishnan, Katherine M Merry, Qiaoqiao Shi, Arnon Rosenthal, and Ben A Barres. 2016. "Complement and Microglia Mediate Early Synapse Loss in Alzheimer Mouse Models." *Science* 352 (6286): 712–16.

Hooks, Bryan M, and Chinfei Chen. 2006. "Distinct Roles for Spontaneous and Visual Activity in Remodeling of the Retinogeniculate Synapse." *Neuron* 52 (2): 281–91.

Horvath, Istvan, Igor A Iashchishyn, Lars Forsgren, and Ludmilla A Morozova-Roche. 2017. "Immunochemical Detection of α-Synuclein Autoantibodies in Parkinson's Disease: Correlation between Plasma and Cerebrospinal Fluid Levels." *ACS Chemical Neuroscience* 8 (6): 1170–76.

Hoshiko, Maki, Isabelle Arnoux, Elena Avignone, Nobuhiko Yamamoto, and Etienne Audinat. 2012. "Deficiency of the Microglial Receptor CX3CR1 Impairs Postnatal Functional Development of Thalamocortical Synapses in the Barrel Cortex." *Journal of Neuroscience* 32 (43): 15106–11.

Hove, Hannah Van, Liesbet Martens, Isabelle Scheyltjens, Karen De Vlaminck, Ana Rita Pombo Antunes, Sofie De Prijck, Niels Vandamme, Sebastiaan De Schepper, Gert Van Isterdael, and Charlotte L Scott. 2019. "A Single-Cell Atlas of Mouse Brain Macrophages Reveals Unique Transcriptional Identities Shaped by Ontogeny and Tissue Environment." *Nature Neuroscience* 22 (6): 1021–35.

Hsiao, Han-Yun, Yu-Chen Chen, Hui-Mei Chen, Pang-Hsien Tu, and Yijuang Chern. 2013. "A Critical Role of Astrocyte-Mediated Nuclear Factor-KB-Dependent Inflammation in Huntington's Disease." *Human Molecular Genetics* 22 (9): 1826–42.

Hu, Dan, Samuele Notarbartolo, Tom Croonenborghs, Bonny Patel, Ron Cialic, Tun-Hsiang Yang, Dominik Aschenbrenner, Karin M Andersson, Marco Gattorno, and Minh Pham. 2017. "Transcriptional Signature of Human Pro-Inflammatory TH17 Cells Identifies Reduced IL10 Gene Expression in Multiple Sclerosis." *Nature Communications* 8 (1): 1600.

Hu, Z X, W N Song, X D Lu, M L Zhou, and J H Shao. 2018. "Peripheral T Lymphocyte Immunity and L-Dopamine in Patients with Parkinson's Disease." *Journal of Biological Regulators and Homeostatic Agents* 32 (3): 687–91.

Huang, Wei, Qiankun Lv, Yunfei Xiao, Zhen Zhong, Binbin Hu, Si Yan, Yufang Yan, Junjun Zhang, Ting Shi, and Lijuan Jiang. 2021. "Triggering Receptor Expressed on Myeloid Cells 2 Protects Dopaminergic Neurons by Promoting Autophagy in the Inflammatory Pathogenesis of Parkinson's Disease." *Frontiers in Neuroscience* 15: 745815.

Hughes, Ethan G, Shin H Kang, Masahiro Fukaya, and Dwight E Bergles. 2013. "Oligodendrocyte Progenitors Balance Growth with Self-Repulsion to Achieve Homeostasis in the Adult Brain." *Nature Neuroscience* 16 (6): 668–76.

Hyman, Bradley T, Carolyn Smith, Ilya Buldyrev, Christa Whelan, Harold Brown, Ming-Xin Tang, and Richard Mayeux. 2001. "Autoantibodies to Amyloid-β and Alzheimer's Disease." *Annals of Neurology: Official Journal of the American Neurological Association and the Child Neurology Society* 49 (6): 808–10.

Investigators, Ninds Net-Pd. 2006. "A Randomized, Double-Blind, Futility Clinical Trial of Creatine and Minocycline in Early Parkinson Disease." *Neurology* 66 (5): 664–71.

Investigators, Ninds Net-Pd. 2008. "A Pilot Clinical Trial of Creatine and Minocycline in Early Parkinson Disease: 18-Month Results." *Clinical Neuropharmacology* 31 (3): 141.

Iram, Tal, Zaida Ramirez-Ortiz, Michael H Byrne, Uwanda A Coleman, Nathan D Kingery, Terry K Means, Dan Frenkel, and Joseph El Khoury. 2016. "Megf10 Is a Receptor for C1Q That Mediates Clearance of Apoptotic Cells by Astrocytes." *Journal of Neuroscience* 36 (19): 5185–92.

Ising, Christina, Carmen Venegas, Shuangshuang Zhang, Hannah Scheiblich, Susanne V Schmidt, Ana Vieira-Saecker, Stephanie Schwartz, Shadi Albasset, Róisín M McManus, and Dario Tejera. 2019. "NLRP3 Inflammasome Activation Drives Tau Pathology." *Nature* 575 (7784): 669–73.

Itagaki, S, P L McGeer, and H Akiyama. 1988. "Presence of T-Cytotoxic Suppressor and Leucocyte Common Antigen Positive Cells in Alzheimer's Disease Brain Tissue." *Neuroscience Letters* 91 (3): 259–64.

Jack Jr, Clifford R, David A Bennett, Kaj Blennow, Maria C Carrillo, Billy Dunn, Samantha Budd Haeberlein, David M Holtzman, William Jagust, Frank Jessen, and Jason Karlawish. 2018. "NIA-AA Research Framework: Toward a Biological Definition of Alzheimer's Disease." *Alzheimer's & Dementia* 14 (4): 535–62.

Janus, Christopher, Jacqueline Pearson, JoAnne McLaurin, Paul M Mathews, Ying Jiang, Stephen D Schmidt, M Azhar Chishti, Patrick Horne, Donna Heslin, and Janet French. 2000. "Aβ Peptide Immunization Reduces Behavioural Impairment and Plaques in a Model of Alzheimer's Disease." *Nature* 408 (6815): 979–82.

Jekabsone, Aiste, Palwinder K Mander, Anna Tickler, Martyn Sharpe, and Guy C Brown. 2006. "Fibrillar Beta-Amyloid Peptide Aβ1–40 Activates Microglial Proliferation via Stimulating TNF-α Release and H2O2 Derived from NADPH Oxidase: A Cell Culture Study." *Journal of Neuroinflammation* 3 (1): 1–13.

Jersild, Casper, Arne Svejgaard, and Torben Fog. 1972. "HL-A Antigens and Multiple Sclerosis." *The Lancet* 299 (7762): 1240–41.

Jiang, Sen, Hua Gao, Qin Luo, Pengfei Wang, and Xinling Yang. 2017. "The Correlation of Lymphocyte Subsets, Natural Killer Cell, and Parkinson's Disease: A Meta-Analysis." *Neurological Sciences* 38 (8): 1373–80.

Johnson, Lincoln V, William P Leitner, Alexander J Rivest, Michelle K Staples, Monte J Radeke, and Don H Anderson. 2002. "The Alzheimer's Aβ-Peptide Is Deposited at Sites of Complement Activation in Pathologic Deposits Associated with Aging and Age-Related Macular Degeneration." *Proceedings of the National Academy of Sciences* 99 (18): 11830–35.

Jordão, Marta Joana Costa, Roman Sankowski, Stefanie M Brendecke, Sagar, Giuseppe Locatelli, Yi-Heng Tai, Tuan Leng Tay, Eva Schramm, Stephan Armbruster, and Nora Hagemeyer. 2019. "Single-Cell Profiling Identifies Myeloid Cell Subsets with Distinct Fates during Neuroinflammation." *Science* 363 (6425): eaat7554.

Kalia, Lorraine V, and Anthony E Lang. 2015. "Parkinson's Disease." *The Lancet* 386 (9996): 896–912.

Kamphuis, Willem, Lieneke Kooijman, Sjoerd Schetters, Marie Orre, and Elly M Hol. 2016. "Transcriptional Profiling of CD11c-Positive Microglia Accumulating around Amyloid Plaques in a Mouse Model for Alzheimer's Disease." *Biochimica et Biophysica Acta (BBA)-Molecular Basis of Disease* 1862 (10): 1847–60.

Kawaguchi, Yoshiya, Toshihiro Okamoto, Masafumi Taniwaki, Megumi Aizawa, Miho Inoue, Sadao Katayama, Hideshi Kawakami, Shigenobu Nakamura, Masaki Nishimura, and Ichiro Akiguchi. 1994. "CAG Expansions in a Novel Gene for Machado-Joseph Disease at Chromosome 14q32. 1." *Nature Genetics* 8 (3): 221–28.

Kazantsev, A, E Preisinger, A Dranovsky, D Goldgaber, and D Housman. 1999. "Insoluble Detergent-Resistant Aggregates Form between Pathological and Nonpathological Lengths of Polyglutamine in Mammalian Cells." *Proceedings of the National Academy of Sciences* 96 (20): 11404–9.

Kettenmann, Helmut, and Bruce R Ransom. 2004. "*The Concept of Neuroglia: A Historical Perspective*." In Helmut Kettenmann and Bruce R. Ransom (eds), *Neuroglia*. Oxford University Press.

Kim, Jong-hyeon, Dong-joo Choi, Hey-kyeong Jeong, Jun Kim, Dae Won Kim, Soo Young Choi, Sang-Myun Park, Young Ho Suh, Ilo Jou, and Eun-Hye Joe. 2013. "DJ-1 Facilitates the Interaction between STAT1 and Its Phosphatase, SHP-1, in Brain Microglia and Astrocytes: A Novel Anti-Inflammatory Function of DJ-1." *Neurobiology of Disease* 60: 1–10.

Kim, Kyung Hwa, Minhwan Kim, Jaehwan Lee, Hat Nim Jeon, Se Hyun Kim, and Hyunsu Bae. 2019. "Comparison of the Protective Effects of Bee Venom Extracts with Varying Pla2 Compositions in a Mouse Model of Parkinson's Disease." *Toxins* 11 (6): 358.

Kivisäkk, Pia, Don J Mahad, Melissa K Callahan, Corinna Trebst, Barbara Tucky, Tao Wei, Lijun Wu, Espen S Baekkevold, Hans Lassmann, and Susan M Staugaitis. 2003. "Human Cerebrospinal Fluid Central Memory CD4+ T Cells: Evidence for Trafficking through Choroid Plexus and Meninges via P-Selectin." *Proceedings of the National Academy of Sciences* 100 (14): 8389–94.

Kobo, Hila, Anat Bar-Shira, Dvir Dahary, Ziv Gan-Or, Anat Mirelman, Orly Goldstein, Nir Giladi, and Avi Orr-Urtreger. 2016. "Down-Regulation of B Cell-Related Genes in Peripheral Blood Leukocytes of Parkinson's Disease Patients with and without GBA Mutations." *Molecular Genetics and Metabolism* 117 (2): 179–85.

Kwan, Wanda, Ulrike Träger, Dimitrios Davalos, Austin Chou, Jill Bouchard, Ralph Andre, Aaron Miller, Andreas Weiss, Flaviano Giorgini, and Christine Cheah. 2012. "Mutant Huntingtin Impairs Immune Cell Migration in Huntington Disease." *The Journal of Clinical Investigation* 122 (12): 4737–47.

Lamine, Asma, Myriam Létourneau, Ngoc Duc Doan, Julie Maucotel, Alain Couvineau, Hubert Vaudry, David Chatenet, David Vaudry, and Alain Fournier. 2016. "Characterizations of a Synthetic Pituitary Adenylate Cyclase-Activating Polypeptide Analog Displaying Potent Neuroprotective Activity and Reduced in Vivo Cardiovascular Side Effects in a Parkinson's Disease Model." *Neuropharmacology* 108: 440–50.

Lang, Heather L E, Helle Jacobsen, Shinji Ikemizu, Christina Andersson, Karl Harlos, Lars Madsen, Peter Hjorth, Leif Sondergaard, Arne Svejgaard, and Kai Wucherpfennig. 2002. "A Functional and Structural Basis for TCR Cross-Reactivity in Multiple Sclerosis." *Nature Immunology* 3 (10): 940–43.

Lau, Lok Ting, and Albert Cheung-Hoi Yu. 2001. "Astrocytes Produce and Release Interleukin-1, Interleukin-6, Tumor Necrosis Factor Alpha and Interferon-Gamma Following Traumatic and Metabolic Injury." *Journal of Neurotrauma* 18 (3): 351–59.

Laurent, Cyril, Guillaume Dorothée, Stéphane Hunot, Elodie Martin, Yann Monnet, Marie Duchamp, Yuan Dong, François-Pierre Légeron, Antoine Leboucher, and Sylvie Burnouf. 2017. "Hippocampal T Cell Infiltration Promotes Neuroinflammation and Cognitive Decline in a Mouse Model of Tauopathy." *Brain* 140 (1): 184–200.

Leal, María C, Juan C Casabona, Mariana Puntel, and Fernando J Pitossi. 2013. "Interleukin-1β and Tumor Necrosis Factor-α: Reliable Targets for Protective Therapies in Parkinson's Disease?" *Frontiers in Cellular Neuroscience* 7: 53.

Lee, C Y Daniel, and Gary E Landreth. 2010. "The Role of Microglia in Amyloid Clearance from the AD Brain." *Journal of Neural Transmission* 117: 949–60.

Lee, Eun-Jung, Moon-Sook Woo, Pyong-Gon Moon, Moon-Chang Baek, In-Young Choi, Won-Ki Kim, Eunsung Junn, and Hee-Sun Kim. 2010a. "α-Synuclein Activates Microglia by Inducing the Expressions of Matrix Metalloproteinases and the Subsequent Activation of Protease-Activated Receptor-1." *The Journal of Immunology* 185 (1): 615–23.

Lee, Sungho, Nicholas H Varvel, Megan E Konerth, Guixiang Xu, Astrid E Cardona, Richard M Ransohoff, and Bruce T Lamb. 2010b. "CX3CR1 Deficiency Alters Microglial Activation and Reduces Beta-Amyloid Deposition in Two Alzheimer's Disease Mouse Models." *The American Journal of Pathology* 177 (5): 2549–62.

Lees, Andrew J, John Hardy, and Tamas Revesz. 2009. "Parkinson's Disease." *The Lancet* 373 (9680): 2055–66.

Levine, Joel M, Richard Reynolds, and James W Fawcett. 2001. "The Oligodendrocyte Precursor Cell in Health and Disease." *Trends in Neurosciences* 24 (1): 39–47.

Lf, Lin. 1993. "GDNF: A Glial Cell Line-Derived Neurotrophic Factor for Midbrain Dopaminergic Neurons." *Science* 260: 1130–32.

Li, Jian, Chunmei Ma, Fei Long, Dongxue Yang, Xue Liu, Yingchao Hu, Chunyan Wu, Bingwei Wang, Min Wang, and Yunzi Chen. 2019a. "Parkin Impairs Antiviral Immunity by Suppressing the Mitochondrial Reactive Oxygen Species-Nlrp3 Axis and Antiviral Inflammation." *Iscience* 16: 468–84.

Li, Xinyi, Wouter Koudstaal, Lauren Fletcher, Martha Costa, Margot van Winsen, Berdien Siregar, Hanna Inganäs, Julie Kim, Elissa Keogh, and Jeremy Macedo. 2019b. "Naturally Occurring Antibodies Isolated from PD Patients Inhibit Synuclein Seeding in Vitro and Recognize Lewy Pathology." *Acta Neuropathologica* 137: 825–36.

Liddelow, Shane A, Kevin A Guttenplan, Laura E Clarke, Frederick C Bennett, Christopher J Bohlen, Lucas Schirmer, Mariko L Bennett, Alexandra E Münch, Won-Suk Chung, and Todd C Peterson. 2017. "Neurotoxic Reactive Astrocytes Are Induced by Activated Microglia." *Nature* 541 (7638): 481–87.

Lindestam Arlehamn, Cecilia S, Rekha Dhanwani, John Pham, Rebecca Kuan, April Frazier, Juliana Rezende Dutra, Elizabeth Phillips, Simon Mallal, Mario Roederer, and Karen S Marder. 2020. "α-Synuclein-Specific T Cell Reactivity Is Associated with Preclinical and Early Parkinson's Disease." *Nature Communications* 11 (1): 1875.

Linnerbauer, Mathias, Michael A Wheeler, and Francisco J Quintana. 2020. "Astrocyte Crosstalk in CNS Inflammation." *Neuron* 108 (4): 608–22.

Liu, Tsai-Wei, Chiung-Mei Chen, and Kuo-Hsuan Chang. 2022. "Biomarker of Neuroinflammation in Parkinson's Disease." *International Journal of Molecular Sciences* 23 (8): 4148.

Liu, Yang, Silke Walter, Massimiliano Stagi, Dmitry Cherny, Maryse Letiembre, Walter Schulz-Schaeffer, Holger Heine, Botond Penke, Harald Neumann, and Klaus Fassbender. 2005. "LPS Receptor (CD14): A Receptor for Phagocytosis of Alzheimer's Amyloid Peptide." *Brain* 128 (8): 1778–89.

Lue, Lih-Fen, Douglas G Walker, and Joseph Rogers. 2001. "Modeling Microglial Activation in Alzheimer's Disease with Human Postmortem Microglial Cultures." *Neurobiology of Aging* 22 (6): 945–56.

Madden, Kelley S, and David L Felten. 1995. "Experimental Basis for Neural-Immune Interactions." *Physiological Reviews* 75 (1): 77–106.

Madetko, Natalia, Bartosz Migda, Piotr Alster, Paweł Turski, Dariusz Koziorowski, and Andrzej Friedman. 2022. "Platelet-to-Lymphocyte Ratio and Neutrophil-Tolymphocyte Ratio May Reflect Differences in PD and MSA-P Neuroinflammation Patterns." *Neurologia i Neurochirurgia Polska* 56 (2): 148–55.

Malagelada, Cristina, Zong Hao Jin, Vernice Jackson-Lewis, Serge Przedborski, and Lloyd A Greene. 2010. "Rapamycin Protects against Neuron Death in in Vitro AndIn Vivo Models of Parkinson's Disease." *Journal of Neuroscience* 30 (3): 1166–75.

Mandler, Markus, Elvira Valera, Edward Rockenstein, Harald Weninger, Christina Patrick, Anthony Adame, Radmila Santic, Stefanie Meindl, Benjamin Vigl, and Oskar Smrzka. 2014. "Next-Generation Active Immunization Approach for Synucleinopathies: Implications for Parkinson's Disease Clinical Trials." *Acta Neuropathologica* 127: 861–79.

Manthripragada, Angelika D, Eva S Schernhammer, Jiaheng Qiu, Soren Friis, Lene Wermuth, Jorgen H Olsen, and Beate Ritz. 2011. "Non-Steroidal Anti-Inflammatory Drug Use and the Risk of Parkinson's Disease." *Neuroepidemiology* 36 (3): 155–61.

Martinez-Forero, Ivan, Ricardo Garcia-Munoz, Sara Martinez-Pasamar, Susana Inoges, Ascensión Lopez-Diaz de Cerio, Ricardo Palacios, Jorge Sepulcre, Beatriz Moreno, Zaira Gonzalez, and Begoña Fernandez-Diez. 2008. "IL-10 Suppressor Activity and Ex Vivo Tr1 Cell Function Are Impaired in Multiple Sclerosis." *European Journal of Immunology* 38 (2): 576–86.

Masliah, Eliezer, Edward Rockenstein, Anthony Adame, Michael Alford, Leslie Crews, Makoto Hashimoto, Peter Seubert, Michael Lee, Jason Goldstein, and Tamie Chilcote. 2005. "Effects of α-Synuclein Immunization in a Mouse Model of Parkinson's Disease." *Neuron* 46 (6): 857–68.

Matsumoto, Yoh, Kuniko Kohyama, Yukihiko Aikawa, Taekyun Shin, Yoko Kawazoe, Yoko Suzuki, and Naoyuki Tanuma. 1998. "Role of Natural Killer Cells and TCRγ δ T Cells in Acute Autoimmune Encephalomyelitis." *European Journal of Immunology* 28 (5): 1681–8.

Mayne, Katie, Jessica A White, Christopher E McMurran, Francisco J Rivera, and Alerie G de la Fuente. 2020. "Aging and Neurodegenerative Disease: Is the Adaptive Immune System a Friend or Foe?" *Frontiers in Aging Neuroscience* 12: 572090.

McGeer, P L, H Akiyama, S Itagaki, and E G McGeer. 1989. "Activation of the Classical Complement Pathway in Brain Tissue of Alzheimer Patients." *Neuroscience Letters* 107 (1–3): 341–46.

McGeer, Patrick L, and Edith G McGeer. 2004. "Inflammation and Neurodegeneration in Parkinson's Disease." *Parkinsonism & Related Disorders* 10: S3–7.

Medana, Isabelle M, Awen Gallimore, Annette Oxenius, Marianne M A Martinic, Hartmut Wekerle, and Harald Neumann. 2000. "MHC Class I-restricted Killing of Neurons by Virus-specific CD8+ T Lymphocytes Is Effected through the Fas/FasL, but Not the Perforin Pathway." *European Journal of Immunology* 30 (12): 3623–33.

Mejia, René O Sanchez, and Robert M Friedlander. 2001. "Caspases in Huntington's Disease." *The Neuroscientist* 7 (6): 480–89.

Mengel, D, S Röskam, F Neff, K Balakrishnan, O Deuster, M Gold, W H Oertel, M Bacher, J P Bach, and R Dodel. 2013. "Naturally Occurring Autoantibodies Interfere with β-Amyloid Metabolism and Improve Cognition in a Transgenic Mouse Model of Alzheimer's Disease 24 h after Single Treatment." *Translational Psychiatry* 3 (3): e236–e236.

Merlini, Mario, Tunahan Kirabali, Luka Kulic, Roger M Nitsch, and Maria Teresa Ferretti. 2018. "Extravascular CD3+ T Cells in Brains of Alzheimer Disease Patients Correlate with Tau but Not with Amyloid Pathology: An Immunohistochemical Study." *Neurodegenerative Diseases* 18 (1): 49–56.

Mintun, Mark A, Albert C Lo, Cynthia Duggan Evans, Alette M Wessels, Paul A Ardayfio, Scott W Andersen, Sergey Shcherbinin, JonDavid Sparks, John R Sims, and Miroslaw Brys. 2021. "Donanemab in Early Alzheimer's Disease." *New England Journal of Medicine* 384 (18): 1691–1704.

Mittal, Kritika, Ekaterina Eremenko, Omer Berner, Yehezqel Elyahu, Itai Strominger, Daniella Apelblat, Anna Nemirovsky, Ivo Spiegel, and Alon Monsonego. 2019. "CD4 T Cells Induce a Subset of MHCII-Expressing Microglia That Attenuates Alzheimer Pathology." *Iscience* 16: 298–311.

Monsonego, Alon, and Howard L Weiner. 2003. "Immunotherapeutic Approaches to Alzheimer's Disease." *Science* 302 (5646): 834–38.

Morgan, Dave, David M Diamond, Paul E Gottschall, Kenneth E Ugen, Chad Dickey, John Hardy, Karen Duff, Paul Jantzen, Giovanni DiCarlo, and Donna Wilcock. 2000. "Aβ Peptide Vaccination Prevents Memory Loss in an Animal Model of Alzheimer's Disease." *Nature* 408 (6815): 982–85.

Morrison, A Norman, M Peterson, F L Raymond, S Simpson, E Thompson, J F Heydon, E Kinning, A Lashwood, M Losekoot, and L Margerison. 2007. "Reduced Penetrance Alleles for Huntington's." *Journal of Medical Genetics* 44: e68.

Mumford, Colin J, Nick W Wood, H Kellar-Wood, John W Thorpe, David H Miller, and D A S Compston. 1994. "The British Isles Survey of Multiple Sclerosis in Twins." *Neurology* 44 (1): 11.

Mundt, Sarah, Melanie Greter, Alexander Flügel, and Burkhard Becher. 2019. "The CNS Immune Landscape from the Viewpoint of a T Cell." *Trends in Neurosciences* 42 (10): 667–79.

Munschauer, F E, L A Hartrich, C C Stewart, and L Jacobs. 1995. "Circulating Natural Killer Cells but Not Cytotoxic T Lymphocytes Are Reduced in Patients with Active Relapsing Multiple Sclerosis and Little Clinical Disability as Compared to Controls." *Journal of Neuroimmunology* 62 (2): 177–81.

Nagele, Robert G, Jerzy Wegiel, Venkat Venkataraman, Humi Imaki, Kuo-Chiang Wang, and Jarek Wegiel. 2004. "Contribution of Glial Cells to the Development of Amyloid Plaques in Alzheimer's Disease." *Neurobiology of Aging* 25 (5): 663–74.

Naito, S, N Namerow, M R Mickey, and P I Terasaki. 1972. "Multiple Sclerosis: Association with HL—A3." *Tissue Antigens* 2 (1): 1–4.

Nathan, Carl, Noel Calingasan, Jon Nezezon, Aihao Ding, M Scott Lucia, Krista La Perle, Michele Fuortes, Michael Lin, Sabine Ehrt, and Nyoun Soo Kwon. 2005. "Protection from Alzheimer's-like Disease in the Mouse by Genetic Ablation of Inducible Nitric Oxide Synthase." *The Journal of Experimental Medicine* 202 (9): 1163–69.

Nave, Klaus-Armin. 2010. "Myelination and Support of Axonal Integrity by Glia." *Nature* 468 (7321): 244–52.

Negi, Neema, and Bimal K Das. 2018. "CNS: Not an Immunoprivilaged Site Anymore but a Virtual Secondary Lymphoid Organ." *International Reviews of Immunology* 37 (1): 57–68.

Nichols, R Jeremy, Nicolas Dzamko, Nicholas A Morrice, David G Campbell, Maria Deak, Alban Ordureau, Thomas Macartney, Youren Tong, Jie Shen, and Alan R Prescott. 2010. "14-3-3 Binding to LRRK2 Is Disrupted by Multiple Parkinson's Disease-Associated Mutations and Regulates Cytoplasmic Localization." *Biochemical Journal* 430 (3): 393–404.

Oksanen, Minna, Sarka Lehtonen, Merja Jaronen, Gundars Goldsteins, Riikka H Hämäläinen, and Jari Koistinaho. 2019. “Astrocyte Alterations in Neurodegenerative Pathologies and Their Modeling in Human Induced Pluripotent Stem Cell Platforms.” *Cellular and Molecular Life Sciences* 76: 2739–60.

Olejniczak, Marta, Martyna O Urbanek, and Wlodzimierz J Krzyzosiak. 2015. “The Role of the Immune System in Triplet Repeat Expansion Diseases.” *Mediators of Inflammation* 2015: 873860.

Olsson, Tomas, Lisa F Barcellos, and Lars Alfredsson. 2017. “Interactions between Genetic, Lifestyle and Environmental Risk Factors for Multiple Sclerosis.” *Nature Reviews Neurology* 13 (1): 25–36.

Ona, Victor O, Mingwei Li, Jean Paul G Vonsattel, L John Andrews, Sohail Q Khan, Woosik M Chung, Ariel S Frey, Anil S Menon, Xiao-Jiang Li, and Philip E Stieg. 1999. “Inhibition of Caspase-1 Slows Disease Progression in a Mouse Model of Huntington’s Disease.” *Nature* 399 (6733): 263–67.

Orr, Carolyn F, Dominic B Rowe, Yoshikuni Mizuno, Hideo Mori, and Glenda M Halliday. 2005. “A Possible Role for Humoral Immunity in the Pathogenesis of Parkinson’s Disease.” *Brain* 128 (11): 2665–74.

Orsini, Franca, Daiana De Blasio, Rosalia Zangari, Elisa R Zanier, and Maria-Grazia De Simoni. 2014. “Versatility of the Complement System in Neuroinflammation, Neurodegeneration and Brain Homeostasis.” *Frontiers in Cellular Neuroscience* 8: 380.

Osellame, Laura D, and Michael R Duchen. 2013. “Defective Quality Control Mechanisms and Accumulation of Damaged Mitochondria Link Gaucher and Parkinson Diseases.” *Autophagy* 9 (10): 1633–35.

Osellame, Laura D, Ahad A Rahim, Iain P Hargreaves, Matthew E Gegg, Angela Richard-Londt, Sebastian Brandner, Simon N Waddington, Anthony H V Schapira, and Michael R Duchen. 2013. “Mitochondria and Quality Control Defects in a Mouse Model of Gaucher Disease—Links to Parkinson’s Disease.” *Cell Metabolism* 17 (6): 941–53.

Ouchi, Yasuomi, Etsuji Yoshikawa, Yoshimoto Sekine, Masami Futatsubashi, Toshihiko Kanno, Tomomi Ogusu, and Tatsuo Torizuka. 2005. “Microglial Activation and Dopamine Terminal Loss in Early Parkinson’s Disease.” *Annals of Neurology* 57 (2): 168–75.

Panicker, Nikhil, Souvarish Sarkar, Dilshan S Harischandra, Matthew Neal, Tae-In Kam, Huajun Jin, Hariharan Saminathan, Monica Langley, Adhithiya Charli, and Manikandan Samidurai. 2019. “Fyn Kinase Regulates Misfolded α-Synuclein Uptake and NLRP3 Inflammasome Activation in Microglia.” *Journal of Experimental Medicine* 216 (6): 1411–30.

Paolicelli, Rosa C, Giulia Bolasco, Francesca Pagani, Laura Maggi, Maria Scianni, Patrizia Panzanelli, Maurizio Giustetto, Tiago Alves Ferreira, Eva Guiducci, and Laura Dumas. 2011. “Synaptic Pruning by Microglia Is Necessary for Normal Brain Development.” *Science* 333 (6048): 1456–58.

Paresce, Donata M, Richik N Ghosh, and Frederick R Maxfield. 1996. “Microglial Cells Internalize Aggregates of the Alzheimer’s Disease Amyloid β-Protein via a Scavenger Receptor.” *Neuron* 17 (3): 553–65.

Parillaud, Vincent R, Guillaume Lornet, Yann Monnet, Anne-Laure Privat, Andrei T Haddad, Vanessa Brochard, Amaury Bekaert, Camille Baudesson de Chanville, Etienne C Hirsch, and Christophe Combadière. 2017. “Analysis of Monocyte Infiltration in MPTP Mice Reveals That Microglial CX3CR1 Protects against Neurotoxic Over-Induction of Monocyte-Attracting CCL2 by Astrocytes.” *Journal of Neuroinflammation* 14 (1): 1–18.

Patel, Nikunj S, Daniel Paris, Venkatarajan Mathura, Amita N Quadros, Fiona C Crawford, and Michael J Mullan. 2005. “Inflammatory Cytokine Levels Correlate with Amyloid Load in Transgenic Mouse Models of Alzheimer’s Disease.” *Journal of Neuroinflammation* 2: 1–10.

Patsopoulos, Nikolaos A, Lisa F Barcellos, Rogier Q Hintzen, Catherine Schaefer, Cornelia M Van Duijn, Janelle A Noble, Towfique Raj, Imsgc, ANZgene, and Pierre-Antoine Gourraud. 2013. "Fine-Mapping the Genetic Association of the Major Histocompatibility Complex in Multiple Sclerosis: HLA and Non-HLA Effects." *PLoS Genetics* 9 (11): e1003926.

Paulson, Henry L, Nancy M Bonini, and Kevin A Roth. 2000. "Polyglutamine Disease and Neuronal Cell Death." *Proceedings of the National Academy of Sciences* 97 (24): 12957–58.

Pavese, N, A Gerhard, Y F Tai, A K Ho, F Turkheimer, R A Barker, D J Brooks, and P Piccini. 2006. "Microglial Activation Correlates with Severity in Huntington Disease: A Clinical and PET Study." *Neurology* 66 (11): 1638–43.

Pekna, Marcela, and Milos Pekny. 2021. "The Complement System: A Powerful Modulator and Effector of Astrocyte Function in the Healthy and Diseased Central Nervous System." *Cells* 10 (7): 1812.

Perez, Matthew K, Henry L Paulson, Sagun J Pendse, Sarah J Saionz, Nancy M Bonini, and Randall N Pittman. 1998. "Recruitment and the Role of Nuclear Localization in Polyglutamine-Mediated Aggregation." *The Journal of Cell Biology* 143 (6): 1457–70.

Perlmutter, L S, S A Scott, E Barron, and H C Chui. 1992. "MHC Class II-Positive Microglia in Human Brain: Association with Alzheimer Lesions." *Journal of Neuroscience Research* 33 (4): 549–58.

Phatnani, Hemali, and Tom Maniatis. 2015. "Astrocytes in Neurodegenerative Disease." *Cold Spring Harbor Perspectives in Biology* 7 (6): a020628.

Pierce, Steven, and Gerhard A Coetzee. 2017. "Parkinson's Disease-Associated Genetic Variation Is Linked to Quantitative Expression of Inflammatory Genes." *PLoS One* 12 (4): e0175882.

Postuma, Ronald B, Dag Aarsland, Paolo Barone, David J Burn, Christopher H Hawkes, Wolfgang Oertel, and Tjalf Ziemssen. 2012. "Identifying Prodromal Parkinson's Disease: Pre-motor Disorders in Parkinson's Disease." *Movement Disorders* 27 (5): 617–26.

Pradhan, Arpit Kumar, Rainer Rupprecht, and Gerhard Rammes. 2024. "Translocator Protein and Neurodegeneration: Insights from Alzheimer's Disease." *Neural Regeneration Research.* doi: 10.4103/NRR.NRR-D-24-00246

Pradhan, Arpit Kumar, Tatjana Neumüller, Claudia Klug, Severin Fuchs, Martin Schlegel, Markus Ballmann, Katharina Johanna Tartler, et al. 2023. "Chronic Administration of XBD173 Ameliorates Cognitive Deficits and Neuropathology via 18 KDa Translocator Protein (TSPO) in a Mouse Model of Alzheimer's Disease." *BioRxiv*, January, 2023.02.23.529740. https://doi.org/10.1101/2023.02.23.529740

Prinz, Marco, Daniel Erny, and Nora Hagemeyer. 2017. "Ontogeny and Homeostasis of CNS Myeloid Cells." *Nature Immunology* 18 (4): 385–92.

Prinz, Marco, Steffen Jung, and Josef Priller. 2019. "Microglia Biology: One Century of Evolving Concepts." *Cell* 179 (2): 292–311.

Qin, Xiao-Yan, Shu-Ping Zhang, Chang Cao, Y Peng Loh, and Yong Cheng. 2016. "Aberrations in Peripheral Inflammatory Cytokine Levels in Parkinson Disease: A Systematic Review and Meta-Analysis." *JAMA Neurology* 73 (11): 1316–24.

Quintana, Francisco J. 2019. "Astrocytes Play a Crucial Role in the Formation and Evolution of MS Lesions–Commentary." *Multiple Sclerosis Journal* 25 (1): 19–20.

Quintana, Francisco J, Soledad Pérez-Sánchez, and Mauricio F Farez. 2014. "Inmunopatología de La Esclerosis Múltiple." *Medicina (Buenos Aires)* 74 (5): 404–10.

Raivich, Gennadij, and Richard Banati. 2004. "Brain Microglia and Blood-Derived Macrophages: Molecular Profiles and Functional Roles in Multiple Sclerosis and Animal Models of Autoimmune Demyelinating Disease." *Brain Research Reviews* 46 (3): 261–81.

Rhodes, K E, G Raivich, and J W Fawcett. 2006. "The Injury Response of Oligodendrocyte Precursor Cells Is Induced by Platelets, Macrophages and Inflammation-Associated Cytokines." *Neuroscience* 140 (1): 87–100.

Ricklin, Daniel, George Hajishengallis, Kun Yang, and John D Lambris. 2010. "Complement: A Key System for Immune Surveillance and Homeostasis." *Nature Immunology* 11 (9): 785–97.

Rivers, Leanne E, Kaylene M Young, Matteo Rizzi, Françoise Jamen, Konstantina Psachoulia, Anna Wade, Nicoletta Kessaris, and William D Richardson. 2008. "PDGFRA/NG2 Glia Generate Myelinating Oligodendrocytes and Piriform Projection Neurons in Adult Mice." *Nature Neuroscience* 11 (12): 1392–1401.

Robertson, N P, M Fraser, J Deans, D Clayton, N Walker, and D A S Compston. 1996. "Age-Adjusted Recurrence Risks for Relatives of Patients with Multiple Sclerosis." *Brain* 119 (2): 449–55.

Rocha, Natália Pessoa, Aline Silva De Miranda, and Antônio Lúcio Teixeira. 2015. "Insights into Neuroinflammation in Parkinson's Disease: From Biomarkers to Anti-Inflammatory Based Therapies." *BioMed Research International* 2015: 628192.

Rosenkranz, Daniela, Sascha Weyer, Eva Tolosa, Alexandra Gaenslen, Daniela Berg, Thomas Leyhe, Thomas Gasser, and Lars Stoltze. 2007. "Higher Frequency of Regulatory T Cells in the Elderly and Increased Suppressive Activity in Neurodegeneration." *Journal of Neuroimmunology* 188 (1–2): 117–27.

Rustenhoven, Justin, Antoine Drieu, Tornike Mamuladze, Kalil Alves de Lima, Taitea Dykstra, Morgan Wall, Zachary Papadopoulos, Mitsuhiro Kanamori, Andrea Francesca Salvador, and Wendy Baker. 2021. "Functional Characterization of the Dural Sinuses as a Neuroimmune Interface." *Cell* 184 (4): 1000–1016.

Sadovnick, A Dessa, Irene M L Yee, George C Ebers, and Neil J Risch. 1998. "Effect of Age at Onset and Parental Disease Status on Sibling Risks for MS." *Neurology* 50 (3): 719–23.

Samii, Ali, Mahyar Etminan, Matthew O Wiens, and Siavash Jafari. 2009. "NSAID Use and the Risk of Parkinson's Disease: Systematic Review and Meta-Analysis of Observational Studies." *Drugs & Aging* 26: 769–79.

Sapp, E, K B Kegel, N Aronin, T Hashikawa, Y Uchiyama, K Tohyama, P G Bhide, J P Vonsattel, and M DiFiglia. 2001. "Early and Progressive Accumulation of Reactive Microglia in the Huntington Disease Brain." *Journal of Neuropathology & Experimental Neurology* 60 (2): 161–72.

Savarin-Vuaillat, Carine, and Richard M Ransohoff. 2007. "Chemokines and Chemokine Receptors in Neurological Disease: Raise, Retain, or Reduce?" *Neurotherapeutics* 4: 590–601.

Schafer, Dorothy P, Emily K Lehrman, Amanda G Kautzman, Ryuta Koyama, Alan R Mardinly, Ryo Yamasaki, Richard M Ransohoff, Michael E Greenberg, Ben A Barres, and Beth Stevens. 2012. "Microglia Sculpt Postnatal Neural Circuits in an Activity and Complement-Dependent Manner." *Neuron* 74 (4): 691–705.

Schafer, Dorothy P, and Beth Stevens. 2015. "Microglia Function in Central Nervous System Development and Plasticity." *Cold Spring Harbor Perspectives in Biology* 7 (10): a020545.

Schartz, Nicole D, and Andrea J Tenner. 2020. "The Good, the Bad, and the Opportunities of the Complement System in Neurodegenerative Disease." *Journal of Neuroinflammation* 17 (1): 1–25.

Schenk, Dale, Robin Barbour, Whitney Dunn, Grace Gordon, Henry Grajeda, Teresa Guido, Kang Hu, Jiping Huang, Kelly Johnson-Wood, and Karen Khan. 1999. "Immunization with Amyloid-β Attenuates Alzheimer-Disease-like Pathology in the PDAPP Mouse." *Nature* 400 (6740): 173–77.

Schlachetzki, Johannes, Iryna Prots, Jenhan Tao, Hyun B Chun, Kaoru Saijo, David Gosselin, Beate Winner, Christopher K Glass, and Jürgen Winkler. 2018. "A Monocyte Gene Expression Signature in the Early Clinical Course of Parkinson's Disease." *Scientific Reports* 8 (1): 1–13.

Schmidt, Thorsten, G Bernhard Landwehrmeyer, Ina Schmitt, Yvon Trottier, Georg Auburger, Franco Laccone, Thomas Klockgether, Michael Völpel, Jörg T Epplen, and Ludger Schöls. 1998. "An Isoform of Ataxin-3 Accumulates in the Nucleus of Neuronal Cells in Affected Brain Regions of SCA3 Patients." *Brain Pathology* 8 (4): 669–79.

Schroder, Kate, and Jurg Tschopp. 2010. "The Inflammasomes." *Cell* 140 (6): 821–32.

Schutt, Charles R, Howard E Gendelman, and R Lee Mosley. 2018. "Tolerogenic Bone Marrow-Derived Dendritic Cells Induce Neuroprotective Regulatory T Cells in a Model of Parkinson's Disease." *Molecular Neurodegeneration* 13 (1): 1–17.

Sevigny, Jeff, Ping Chiao, Thierry Bussière, Paul H Weinreb, Leslie Williams, Marcel Maier, Robert Dunstan, Stephen Salloway, Tianle Chen, and Yan Ling. 2016. "The Antibody Aducanumab Reduces Aβ Plaques in Alzheimer's Disease." *Nature* 537 (7618): 50–56.

Shaftel, Solomon S, Thaddeus J Carlson, John A Olschowka, Stephanos Kyrkanides, Sarah B Matousek, and M Kerry O'Banion. 2007. "Chronic Interleukin-1β Expression in Mouse Brain Leads to Leukocyte Infiltration and Neutrophil-Independent Blood–Brain Barrier Permeability without Overt Neurodegeneration." *Journal of Neuroscience* 27 (35): 9301–9.

Shah, Akash, Uday Kishore, and Abhishek Shastri. 2021. "Complement System in Alzheimer's Disease." *International Journal of Molecular Sciences* 22 (24): 13647.

Sierra, Amanda, Juan M Encinas, Juan J P Deudero, Jessica H Chancey, Grigori Enikolopov, Linda S Overstreet-Wadiche, Stella E Tsirka, and Mirjana Maletic-Savatic. 2010. "Microglia Shape Adult Hippocampal Neurogenesis through Apoptosis-Coupled Phagocytosis." *Cell Stem Cell* 7 (4): 483–95.

Singhrao, S K, J W Neal, B P Morgan, and P Gasque. 1999. "Increased Complement Biosynthesis by Microglia and Complement Activation on Neurons in Huntington's Disease." *Experimental Neurology* 159 (2): 362–76.

Smits, Hessel A, Annemarie Rijsmus, Joyce H van Loon, Jesse W Y Wat, Jan Verhoef, Leonie A Boven, and Hans S L M Nottet. 2002. "Amyloid-β-Induced Chemokine Production in Primary Human Macrophages and Astrocytes." *Journal of Neuroimmunology* 127 (1–2): 160–8.

Sofroniew, Michael V. 2009. "Molecular Dissection of Reactive Astrogliosis and Glial Scar Formation." *Trends in Neurosciences* 32 (12): 638–47.

Sofroniew, Michael V, and Harry V Vinters. 2010. "Astrocytes: Biology and Pathology." *Acta Neuropathologica* 119: 7–35.

Solano, Steven M, David W Miller, Sarah J Augood, Anne B Young, and John B Penney Jr. 2000. "Expression of A-synuclein, Parkin, and Ubiquitin Carboxy-terminal Hydrolase L1 MRNA in Human Brain: Genes Associated with Familial Parkinson's Disease." *Annals of Neurology: Official Journal of the American Neurological Association and the Child Neurology Society* 47 (2): 201–10.

Soldano, Stefano, Amelia Chiara Trombetta, Paola Contini, Veronica Tomatis, Barbara Ruaro, Renata Brizzolara, Paola Montagna, Alberto Sulli, Sabrina Paolino, and Carmen Pizzorni. 2018. "Increase in Circulating Cells Coexpressing M1 and M2 Macrophage Surface Markers in Patients with Systemic Sclerosis." *Annals of the Rheumatic Diseases* 77 (12): 1842–45.

Söllvander, Sofia, Frida Ekholm-Pettersson, Rose-Marie Brundin, Gabriel Westman, Lena Kilander, Staffan Paulie, Lars Lannfelt, and Dag Sehlin. 2015. "Increased Number of Plasma B Cells Producing Autoantibodies against Aβ 42 Protofibrils in Alzheimer's Disease." *Journal of Alzheimer's Disease* 48 (1): 63–72.

Sospedra, Mireia, and Roland Martin. 2005. "Immunology of Multiple Sclerosis." *Annu. Rev. Immunol.* 23: 683–747.

Späni, Claudia, Tobias Suter, Rebecca Derungs, Maria Teresa Ferretti, Tobias Welt, Fabian Wirth, Christoph Gericke, Roger M Nitsch, and Luka Kulic. 2015. "Reduced β-Amyloid Pathology in an APP Transgenic Mouse Model of Alzheimer's Disease Lacking Functional B and T Cells." *Acta Neuropathologica Communications* 3 (1): 1–14.

Stephan, Alexander H, Ben A Barres, and Beth Stevens. 2012. "The Complement System: An Unexpected Role in Synaptic Pruning during Development and Disease." *Annual Review of Neuroscience* 35: 369–89.

Stevens, Beth, Nicola J Allen, Luis E Vazquez, Gareth R Howell, Karen S Christopherson, Navid Nouri, Kristina D Micheva, Adrienne K Mehalow, Andrew D Huberman, and Benjamin Stafford. 2007. "The Classical Complement Cascade Mediates CNS Synapse Elimination." *Cell* 131 (6): 1164–78.

Stevens, Claire H, Dominic Rowe, Marie-Christine Morel-Kopp, Carolyn Orr, Tonia Russell, Madelaine Ranola, Christopher Ward, and Glenda M Halliday. 2012. "Reduced T Helper and B Lymphocytes in Parkinson's Disease." *Journal of Neuroimmunology* 252 (1–2): 95–99.

Stewart, Cameron R, Lynda M Stuart, Kim Wilkinson, Janine M Van Gils, Jiusheng Deng, Annett Halle, Katey J Rayner, Laurent Boyer, Ruiqin Zhong, and William A Frazier. 2010. "CD36 Ligands Promote Sterile Inflammation through Assembly of a Toll-like Receptor 4 and 6 Heterodimer." *Nature Immunology* 11 (2): 155–61.

Strohmeyer, Ron, Mauricio Ramirez, Gregory J Cole, Kyle Mueller, and Joseph Rogers. 2002. "Association of Factor H of the Alternative Pathway of Complement with Agrin and Complement Receptor 3 in the Alzheimer's Disease Brain." *Journal of Neuroimmunology* 131 (1–2): 135–46.

Subbarayan, Meena S, Charles Hudson, Lauren D Moss, Kevin R Nash, and Paula C Bickford. 2020. "T Cell Infiltration and Upregulation of MHCII in Microglia Leads to Accelerated Neuronal Loss in an α-Synuclein Rat Model of Parkinson's Disease." *Journal of Neuroinflammation* 17 (1): 1–16.

Sulzer, David, Roy N Alcalay, Francesca Garretti, Lucien Cote, Ellen Kanter, Julian Agin-Liebes, Christopher Liong, Curtis McMurtrey, William H Hildebrand, and Xiaobo Mao. 2017. "T Cells from Patients with Parkinson's Disease Recognize α-Synuclein Peptides." *Nature* 546 (7660): 656–61.

Surmeier, D James, José A Obeso, and Glenda M Halliday. 2017. "Selective Neuronal Vulnerability in Parkinson Disease." *Nature Reviews Neuroscience* 18 (2): 101–13.

Tabrizi, S J, J Workman, P E Hart, L Mangiarini, A Mahal, G Bates, J M Cooper, and A H V Schapira. 2000. "Mitochondrial Dysfunction and Free Radical Damage in the Huntington R6/2 Transgenic Mouse." *Annals of Neurology* 47 (1): 80–86.

Tai, Yen F, Nicola Pavese, Alexander Gerhard, Sarah J Tabrizi, Roger A Barker, David J Brooks, and Paola Piccini. 2007. "Microglial Activation in Presymptomatic Huntington's Disease Gene Carriers." *Brain* 130 (7): 1759–66.

Tang, Y, T Li, J Li, J Yang, H Liu, X J Zhang, and Weidong Le. 2014. "Jmjd3 Is Essential for the Epigenetic Modulation of Microglia Phenotypes in the Immune Pathogenesis of Parkinson's Disease." *Cell Death & Differentiation* 21 (3): 369–80.

Tang, Yu, and Weidong Le. 2016. "Differential Roles of M1 and M2 Microglia in Neurodegenerative Diseases." *Molecular Neurobiology* 53: 1181–94.

Taylor, J Paul, John Hardy, and Kenneth H Fischbeck. 2002. "Toxic Proteins in Neurodegenerative Disease." *Science* 296 (5575): 1991–95.

Teismann, Peter, Kim Tieu, Oren Cohen, Dong-Kug Choi, Du Chu Wu, Daniel Marks, Miquel Vila, Vernice Jackson-Lewis, and Serge Przedborski. 2003. "Pathogenic Role of Glial Cells in Parkinson's Disease." *Movement Disorders: Official Journal of the Movement Disorder Society* 18 (2): 121–29.

Tenner, Andrea J. 2020. "Complement-Mediated Events in Alzheimer's Disease: Mechanisms and Potential Therapeutic Targets." *The Journal of Immunology* 204 (2): 306–15.

The International Multiple Sclerosis Genetics Consortium. "Class II HLA Interactions Modulate Genetic Risk for Multiple Sclerosis." 2015. *Nature Genetics* 47 (10): 1107–13.

Thompson, Alan J, Brenda L Banwell, Frederik Barkhof, William M Carroll, Timothy Coetzee, Giancarlo Comi, Jorge Correale, Franz Fazekas, Massimo Filippi, and Mark S Freedman. 2018a. "Diagnosis of Multiple Sclerosis: 2017 Revisions of the McDonald Criteria." *The Lancet Neurology* 17 (2): 162–73.

Thompson, Alan J, Sergio E Baranzini, Jeroen Geurts, Bernhard Hemmer, and Olga Ciccarelli. 2018b. "Multiple Sclerosis." *The Lancet* 391 (10130): 1622–36.

Togo, Takashi, Haruhiko Akiyama, Eizo Iseki, Hiromi Kondo, Kenji Ikeda, Masanori Kato, Tatsuro Oda, Kuniaki Tsuchiya, and Kenji Kosaka. 2002. "Occurrence of T Cells in the Brain of Alzheimer's Disease and Other Neurological Diseases." *Journal of Neuroimmunology* 124 (1–2): 83–92.

Tolar, Martin, John Hey, Aidan Power, and Susan Abushakra. 2021. "Neurotoxic Soluble Amyloid Oligomers Drive Alzheimer's Pathogenesis and Represent a Clinically Validated Target for Slowing Disease Progression." *International Journal of Molecular Sciences* 22 (12): 6355.

Träger, Ulrike, Ralph Andre, Nayana Lahiri, Anna Magnusson-Lind, Andreas Weiss, Stephan Grueninger, Chris McKinnon, Eva Sirinathsinghji, Shira Kahlon, and Edith L Pfister. 2014. "HTT-Lowering Reverses Huntington's Disease Immune Dysfunction Caused by NFκB Pathway Dysregulation." *Brain* 137 (3): 819–33.

Triebel, F, and T Hercend. 1989. "Subpopulations of Human Peripheral T Gamma Delta Lymphocytes." *Immunology Today* 10 (6): 186–88.

Tsuboi, Yand, and T Yamada. 1994. "Increased Concentration of C4d Complement Protein in CSF in Amyotrophic Lateral Sclerosis." *Journal of Neurology, Neurosurgery & Psychiatry* 57 (7): 859–61.

Turnbull, Andrew V, and Catherine L Rivier. 1999. "Regulation of the Hypothalamic-Pituitary-Adrenal Axis by Cytokines: Actions and Mechanisms of Action." *Physiological Reviews* 79 (1): 1–71.

Ulrich, Jason D, Mary Beth Finn, Yaming Wang, Alice Shen, Thomas E Mahan, Hong Jiang, Floy R Stewart, Laura Piccio, Marco Colonna, and David M Holtzman. 2014. "Altered Microglial Response to Aβ Plaques in APPPS1-21 Mice Heterozygous for TREM2." *Molecular Neurodegeneration* 9 (1): 1–9.

Veerdonk, Frank L Van de, Mihai G Netea, Charles A Dinarello, and Leo A B Joosten. 2011. "Inflammasome Activation and IL-1β and IL-18 Processing during Infection." *Trends in Immunology* 32 (3): 110–16.

Venken, Koen, Niels Hellings, Tom Broekmans, Karen Hensen, Jean-Luc Rummens, and Piet Stinissen. 2008. "Natural Naive CD4+ CD25+ CD127low Regulatory T Cell (Treg) Development and Function Are Disturbed in Multiple Sclerosis Patients: Recovery of Memory Treg Homeostasis during Disease Progression." *The Journal of Immunology* 180 (9): 6411–20.

Vodovotz, Yoram, M Scott Lucia, Kathleen C Flanders, Louis Chesler, Qiao-Wen Xie, Thomas W Smith, Jeffrey Weidner, Richard Mumford, Robert Webber, and Carl Nathan. 1996. "Inducible Nitric Oxide Synthase in Tangle-Bearing Neurons of Patients with Alzheimer's Disease." *The Journal of Experimental Medicine* 184 (4): 1425–33.

Volc, Dieter, Werner Poewe, Alexandra Kutzelnigg, Petra Lührs, Caroline Thun-Hohenstein, Achim Schneeberger, Gergana Galabova, Nour Majbour, Nishant Vaikath, and Omar El-Agnaf. 2020. "Safety and Immunogenicity of the α-Synuclein Active Immunotherapeutic PD01A in Patients with Parkinson's Disease: A Randomised, Single-Blinded, Phase 1 Trial." *The Lancet Neurology* 19 (7): 591–600.

Waak, Jens, Stephanie S Weber, Andrea Waldenmaier, Karin Görner, Marianna Alunni-Fabbroni, Heinrich Schell, Daniela Vogt-Weisenhorn, Thu-Trang Pham, Veerle Reumers, and Veerle Baekelandt. 2009. "Regulation of Astrocyte Inflammatory Responses by the Parkinson's Disease-associated Gene DJ–1." *The FASEB Journal* 23 (8): 2478–89.

Wakade, Chandramohan, Banabihari Giri, Aneeq Malik, Hesam Khodadadi, John C Morgan, Raymond K Chong, and Babak Baban. 2018. "Niacin Modulates Macrophage Polarization in Parkinson's Disease." *Journal of Neuroimmunology* 320: 76–79.

Wallin, Mitchell T, William J Culpepper, Parisa Coffman, Sarah Pulaski, Heidi Maloni, Clare M Mahan, Jodie K Haselkorn, John F Kurtzke, and Veterans Affairs Multiple Sclerosis Centres of Excellence Epidemiology Group. 2012. "The Gulf War Era Multiple Sclerosis Cohort: Age and Incidence Rates by Race, Sex and Service." *Brain* 135 (6): 1778–85.

Wang, Anni, and Bei Ping He. 2009. "Characteristics and Functions of NG2 Cells in Normal Brain and Neuropathology." *Neurological Research* 31 (2): 144–50.

Wang, Jian-Yong, Qing-Qing Zhuang, Lan-Bing Zhu, Hui Zhu, Ting Li, Rui Li, Song-Fang Chen, Chen-Ping Huang, Xiong Zhang, and Jian-Hong Zhu. 2016. "Meta-Analysis of Brain Iron Levels of Parkinson's Disease Patients Determined by Postmortem and MRI Measurements." *Scientific Reports* 6 (1): 36669.

Wang, Jian, Ivan Jelcic, Lena Mühlenbruch, Veronika Haunerdinger, Nora C Toussaint, Yingdong Zhao, Carolina Cruciani, Wolfgang Faigle, Reza Naghavian, and Magdalena Foege. 2020a. "HLA-DR15 Molecules Jointly Shape an Autoreactive T Cell Repertoire in Multiple Sclerosis." *Cell* 183 (5): 1264–81.

Wang, Qinqin, Hongmei Yao, Wenyan Liu, Bailiu Ya, Hongju Cheng, Zhenkai Xing, and Yili Wu. 2021. "Microglia Polarization in Alzheimer's Disease: Mechanisms and a Potential Therapeutic Target." *Frontiers in Aging Neuroscience* 13: 772717.

Wang, Shoutang, Meer Mustafa, Carla M Yuede, Santiago Viveros Salazar, Philip Kong, Hua Long, Michael Ward, Omer Siddiqui, Robert Paul, and Susan Gilfillan. 2020b. "Anti-Human TREM2 Induces Microglia Proliferation and Reduces Pathology in an Alzheimer's Disease Model." *Journal of Experimental Medicine* 217 (9): e20200785.

Warwick, Charles A, Alex L Keyes, Trent M Woodruff, and Yuriy M Usachev. 2021. "The Complement Cascade in the Regulation of Neuroinflammation, Nociceptive Sensitization, and Pain." *Journal of Biological Chemistry* 297 (3): 101085.

Weiss, Andreas, Ulrike Träger, Edward J Wild, Stephan Grueninger, Ruth Farmer, Christian Landles, Rachael I Scahill, Nayana Lahiri, Salman Haider, and Douglas Macdonald. 2012. "Mutant Huntingtin Fragmentation in Immune Cells Tracks Huntington's Disease Progression." *The Journal of Clinical Investigation* 122 (10): 3731–36.

Wellington, Cheryl L, Lisa M Ellerby, Claire-Anne Gutekunst, Danny Rogers, Simon Warby, Rona K Graham, Odell Loubser, Jeremy van Raamsdonk, Yu-Zhou Yang, and Juliette Gafni. 2002. "Caspase Cleavage of Mutant Huntingtin Precedes Neurodegeneration in Huntington's Disease." *Journal of Neuroscience* 22 (18): 7862–72.

Wellington, Cheryl L, Lisa M Ellerby, Abigail S Hackam, Russell L Margolis, Mark A Trifiro, Krista McCutcheon, Guy S Salvesen, Stephanie S Propp, Michael Bromm, and Kathleen J Rowland. 1998. "Caspase Cleavage of Gene Products Associated with Triplet Expansion Disorders Generates Truncated Fragments Containing the Polyglutamine Tract." *Journal of Biological Chemistry* 273 (15): 9158–67.

Wheeler, Michael A, and Francisco J Quintana. 2019. "Regulation of Astrocyte Functions in Multiple Sclerosis." *Cold Spring Harbor Perspectives in Medicine* 9 (1): a029009.

Wild, Edward, Anna Magnusson, Nayana Lahiri, Ulrika Krus, Michael Orth, Sarah J Tabrizi, and Maria Björkqvist. 2011. "Abnormal Peripheral Chemokine Profile in Huntington's Disease." *PLoS Currents* 3: RRN1231.

Wilhelmsson, Ulrika, Lizhen Li, Marcela Pekna, Claes-Henric Berthold, Sofia Blom, Camilla Eliasson, Oliver Renner, Eric Bushong, Mark Ellisman, and Todd E Morgan. 2004. "Absence of Glial Fibrillary Acidic Protein and Vimentin Prevents Hypertrophy of Astrocytic Processes and Improves Post-Traumatic Regeneration." *Journal of Neuroscience* 24 (21): 5016–21.

Willer, C J, D A Dyment, N J Risch, A D Sadovnick, G C Ebers, and §§ Canadian Collaborative Study Group††. 2003. "Twin Concordance and Sibling Recurrence Rates in Multiple Sclerosis." *Proceedings of the National Academy of Sciences* 100 (22): 12877–82.

Williams, Gregory P, Aubrey M Schonhoff, Asta Jurkuvenaite, Nicole J Gallups, David G Standaert, and Ashley S Harms. 2021. "CD4 T Cells Mediate Brain Inflammation and Neurodegeneration in a Mouse Model of Parkinson's Disease." *Brain* 144 (7): 2047–59.

Williams, Gregory P, Aubrey M Schonhoff, Alessandro Sette, and Cecilia S Lindestam Arlehamn. 2022. "Central and Peripheral Inflammation: Connecting the Immune Responses of Parkinson's Disease." *Journal of Parkinson's Disease* 12 (s1): S129–36.

Xiao, Jia-Jia, Ming Yin, Ze-Jian Wang, and Xiao-Ping Wang. 2015. "Transplanted Neural Stem Cells: Playing a Neuroprotective Role by Ceruloplasmin in the Substantia Nigra of PD Model Rats?" *Oxidative Medicine and Cellular Longevity* 2015: 618631.

Xiao, S, C Xu, and J N Jarvis. 2001. "C1q-Bearing Immune Complexes Induce IL-8 Secretion in Human Umbilical Vein Endothelial Cells (HUVEC) through Protein Tyrosine Kinase- and Mitogen-Activated Protein Kinase-Dependent Mechanisms: Evidence That the 126 KD Phagocytic C1q Receptor Mediates Immun." *Clinical & Experimental Immunology* 125 (3): 360–67.

Yamasaki, Ryo, Haiyan Lu, Oleg Butovsky, Nobuhiko Ohno, Anna M Rietsch, Ron Cialic, Pauline M Wu, Camille E Doykan, Jessica Lin, and Anne C Cotleur. 2014. "Differential Roles of Microglia and Monocytes in the Inflamed Central Nervous System." *Journal of Experimental Medicine* 211 (8): 1533–49.

Yan, Aijuan, Yu Zhang, Jingya Lin, Lu Song, Xijin Wang, and Zhenguo Liu. 2018. "Partial Depletion of Peripheral M1 Macrophages Reverses Motor Deficits in MPTP-Treated Mouse by Suppressing Neuroinflammation and Dopaminergic Neurodegeneration." *Frontiers in Aging Neuroscience* 10: 160.

Yanamandra, Kiran, Marina A Gruden, Vida Casaite, Rolandas Meskys, Lars Forsgren, and Ludmilla A Morozova-Roche. 2011. "α-Synuclein Reactive Antibodies as Diagnostic Biomarkers in Blood Sera of Parkinson's Disease Patients." *PloS One* 6 (4): e18513.

Yang, Junhua, Hongbin Yang, Yali Liu, Xia Li, Liming Qin, Huifang Lou, Shumin Duan, and Hao Wang. 2016. "Astrocytes Contribute to Synapse Elimination via Type 2 Inositol 1, 4, 5-Trisphosphate Receptor-Dependent Release of ATP." *Elife* 5: e15043.

Yeh, Felix L, Yuanyuan Wang, Irene Tom, Lino C Gonzalez, and Morgan Sheng. 2016. "TREM2 Binds to Apolipoproteins, Including APOE and CLU/APOJ, and Thereby Facilitates Uptake of Amyloid-Beta by Microglia." *Neuron* 91 (2): 328–40.

Yin, Zhuoran, Divya Raj, Nasrin Saiepour, Debby Van Dam, Nieske Brouwer, Inge R Holtman, Bart J L Eggen, Thomas Möller, Joseph A Tamm, and Aicha Abdourahman. 2017. "Immune Hyperreactivity of Aβ Plaque-Associated Microglia in Alzheimer's Disease." *Neurobiology of Aging* 55: 115–22.

Yuan, Jiaxin, Hanghang Liu, Hao Zhang, Tingting Wang, Qing Zheng, and Zhen Li. 2022. "Controlled Activation of TRPV1 Channels on Microglia to Boost Their Autophagy for Clearance of Alpha-synuclein and Enhance Therapy of Parkinson's Disease." *Advanced Materials* 34 (11): 2108435.

Yun, Seung Pil, Tae-In Kam, Nikhil Panicker, SangMin Kim, Yumin Oh, Jong-Sung Park, Seung-Hwan Kwon, Yong Joo Park, Senthilkumar S Karuppagounder, and Hyejin Park. 2018. "Block of A1 Astrocyte Conversion by Microglia Is Neuroprotective in Models of Parkinson's Disease." *Nature Medicine* 24 (7): 931–38.

Zarobkiewicz, Michał K, Wioleta Kowalska, Jacek Roliński, and Agnieszka A Bojarska-Junak. 2019. "Γδ T Lymphocytes in the Pathogenesis of Multiple Sclerosis and Experimental Autoimmune Encephalomyelitis." *Journal of Neuroimmunology* 330: 67–73.

Zhang, Ben-ning, Takashi Yamamura, Takayuki Kondo, Michio Fujiwara, and Takeshi Tabira. 1997. "Regulation of Experimental Autoimmune Encephalomyelitis by Natural Killer (NK) Cells." *The Journal of Experimental Medicine* 186 (10): 1677–87.

Zhang, Shiqing, Ruili Feng, Yanhui Li, Linhua Gan, Fangfang Zhou, Shiquan Meng, Qian Li, and Tieqiao Wen. 2018. "Degradation of Alpha-Synuclein by Dendritic Cell Factor 1 Delays Neurodegeneration and Extends Lifespan in Drosophila." *Neurobiology of Aging* 67: 67–74.

Zhang, Ye, Kenian Chen, Steven A Sloan, Mariko L Bennett, Anja R Scholze, Sean O'Keeffe, Hemali P Phatnani, Paolo Guarnieri, Christine Caneda, and Nadine Ruderisch. 2014. "An RNA-Sequencing Transcriptome and Splicing Database of Glia, Neurons, and Vascular Cells of the Cerebral Cortex." *Journal of Neuroscience* 34 (36): 11929–47.

Zhao, Ye, Shikara Keshiya, Farzaneh Atashrazm, Jianqun Gao, Lars M Ittner, Dario R Alessi, Glenda M Halliday, Yuhong Fu, and Nicolas Dzamko. 2018. "Nigrostriatal Pathology with Reduced Astrocytes in LRRK2 S910/S935 Phosphorylation Deficient Knockin Mice." *Neurobiology of Disease* 120: 76–87.

Ziv, Yaniv, Noga Ron, Oleg Butovsky, Gennady Landa, Einav Sudai, Nadav Greenberg, Hagit Cohen, Jonathan Kipnis, and Michal Schwartz. 2006. "Immune Cells Contribute to the Maintenance of Neurogenesis and Spatial Learning Abilities in Adulthood." *Nature Neuroscience* 9 (2): 268–75.

3 Do You Have the Nerves to Fight Infection?

Ashwin Kumar Jainarayanan and Ranjeet Singh Mahla
The University of Oxford, Oxford, UK

Mirudula Elanchezhian
Indian Institute of Science Education and Research, Mohali, India

3.1 INTRODUCTION

Neuroscience and immunity have been regarded as two distinct domains. The neuronal system is known to stay separated apart from the immune system by the blood-brain barrier (BBB) system. However, two systems can communicate via secreted signalling cues, the neurotransmitters from the neuron and cytokines chemokines from immune cells, are known to modulate cross-functioning neuronal and immune systems at the juncture (Neuroimmune communication, n.d.). The primary stress pathways investigated were the hypothalamic-pituitary-adrenal (HPA) axis, which, when activated, leads to the synthesis of cortisol or corticosterone, and the sympathetic nervous system, which, upon activation, triggers the release of epinephrine and norepinephrine (Figure 3.1). Enhancing the immunosuppressive effects of cyclophosphamide in rats may be achievable by administering a saccharin solution prior to exposing them to the adverse side effects of cyclophosphamide (Ader and Cohen 1975). This pivotal experiment demonstrated the existence of a complex network of bidirectional interactions between the immune system and CNS in an indirect manner. Neuroanatomists have depicted the sympathetic innervation of both primary and secondary lymphoid organs, while neuroendocrinologists have explored the expression and roles of neuroendocrine receptors on immune cells. Additionally, cell biologists have demonstrated that activated immunocytes release signaling molecules and molecular factors previously identified in the neuroendocrine system, such as adrenocorticotrophin, derived from proopiomelanocortin. The emphasis on understanding brain-to-immune communication channels, rather than immune-to-brain pathways, stems largely from the active involvement of neuroendocrinologists in this field. Their contributions have significantly advanced our knowledge of the chemical nature and receptor mechanisms of major neuroendocrine factors (Guillemin 1978).

The synthesis and roles of neuroendocrine peptide hormones, the influence of these neuroendocrine hormones on the immune system, the sympathetic nervous system's

 DOI: 10.1201/9781003295471-3

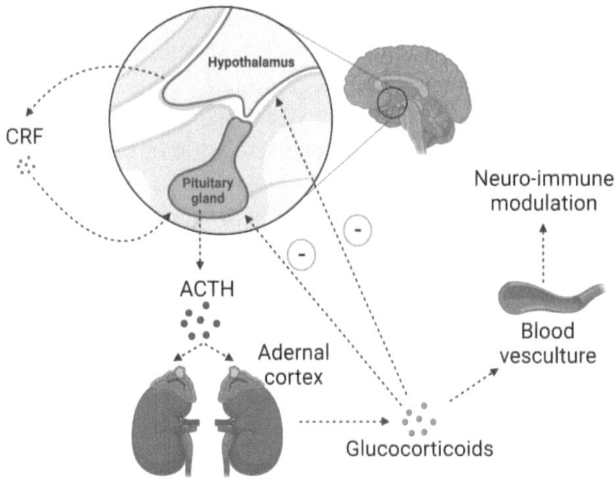

FIGURE 3.1 HPA-axis: Hypothalamus, pituitary, adrenal cortex act in a coordinated manner, regulating secretion of corticosteroids. Release of corticotropin releasing factor (CRF) from the hypothalamus modulates the function of the pituitary gland and release of adrenocorticotropic hormone (ACTH) from the pituitary, further modulating the function of the adrenal cortex and secretion of corticosteroids, which in turn modulate the function of neuro-immune regulation of the metabolism, cardiac function, and much more.

involvement in innervating lymphoid organs (Madden & Felten, 1995), and the regulatory impact of cytokines on the HPA axis (Turnbull & Rivier, 1999) are fundamental communication mechanisms linking the immune system with the nervous system.

3.2 GLIA IN THE CNS AS INNATE IMMUNE CELLS

The CNS is somewhat segregated from the peripheral nervous system, possessing its distinct resident immune network primarily comprised of glial cells, such as microglia and astrocytes. The CNS immune network has two prominent roles, nourishment to support the nervous system and protection against inflammation (Bailey et al. 2006). The maintenance of CNS homeostasis typically relies on various endogenous anti-inflammatory and neuroprotective mechanisms. Prolonged activation of glial cells, however, can lead to severe and chronic neuronal damage, ultimately resulting in neuroinflammation and degeneration. This underscores the intricate nature of regulating neuroprotection and neurotoxicity (Hohlfeld, Kerschensteiner, and Meinl 2007). Similar to chronic neurodegenerative disorders, the microglia in an aging yet otherwise healthy brain tend to be more reactive compared to younger individuals. In this primed state, when stimulated, these microglia release an excessive amount of pro-inflammatory cytokines, thereby exacerbating degenerative processes (Perry 2004). It is important to highlight that glial cells possess the ability to detect immunological responses in the periphery through cytokines and chemokines. This activation alerts the CNS through humoral and/or neurochemical pathways. Subsequently, glial cells initiate the release of a diverse array of chemokines and cytokines, contributing to the manifestation or intensification of sickness behaviors or similar depressive symptoms. This underscores the active communication between central glial cells and the peripheral immune system. In a state of homeostasis, both peripherally and centrally produced inflammatory proteins play crucial roles in guiding adaptive responses of the CNS to both physical and emotional stress (Black P. H., 2002). However, in cases of chronic immune activation or in individuals susceptible to specific neurological disorders, such as schizophrenia, epilepsy, or autism, these inflammatory proteins can have adverse effects on neuronal growth and plasticity. Individuals with such conditions may not tolerate immune stimulation as effectively as those in good health (Carpentier & Palmer, 2009). Broadly glial cells are grouped into oligodendrocytes, microglia, and astrocytes.

3.2.1 Microglia

Microglia constitute 5–20% of all glial cells in various regions of the brain. Within the CNS, microglia serve as the primary phagocytic cells, acting as the first line of defense against pathogenic threats and physiological stress (Rivest, 2009). Playing a pivotal and arguably indispensable role, microglia are directly involved in monitoring and modifying neuronal processes related to neuroimmune crosstalk. In response to inflammation challenges, microglia undergo a transformation into ameboid cells and up-regulate various cell surface receptors associated with innate immune responses. Notable receptors in this context include pattern recognition receptors (PRR) like toll-like receptors (TLR), receptors for advanced glycation end products (RAGE), as well as phagocytic receptors and triggering receptors expressed on myeloid cells (TREM) (Tambuyzer et al., 2009). Phagocytosis, a critical mechanism employed by microglia, plays a vital role in regulating neuronal death. During processes such as apoptosis in neurons (occurring in development, dementia, or injuries), the timely clearance of apoptotic cell debris through phagocytosis is essential to safeguard healthy neurons from inflammation-induced collateral damage. Inadequate clearance of cellular debris through phagocytosis following neuronal injury emerges as a crucial pathogenic element contributing to the advancement of neurodegeneration, a conclusion supported by numerous studies on Alzheimer's disease (AD) (Napoli & Neumann, 2009). Following exposure to Aβ plaques, Microglia may exacerbate the secretion of complement proteins by secreting cytokines such as IL1, IL6, and tumor necrosis factor (TNF)-α. complements and growth factors that have been linked to synapse development and plasticity, in addition to managing neuronal apoptosis. Under both developmental and pathological situations, undesirable synapses are efficiently removed by phagocytosis with the support of the complement system (Bessis et al. 2007). The presentation of foreign antigens to T cells is another crucial function of microglia. Antigen-presenting cells; dendritic cells, and macrophages are predominantly present in meninges, choroid plexuses, and CNS perivascular spaces (Becher, Bechmann, and Greter 2006).

3.2.2 Astrocyte

Traditionally, astrocytes have been considered supportive cells for neurons, playing a crucial role in maintaining CNS homeostasis and facilitates neuronal functions.

The glia limitans are formed by astrocytes around blood arteries, preventing immune cells from crossing the BBB and entering the CNS parenchyma. Astrocytes are known to express all the receptors of innate immune functions including complement receptors, mannose receptors, scavenger receptors, and TLRs (Farina, Aloisi, and Meinl 2007). Following the process of PRRs, astrocytes release chemokines, cytokines and neurotrophins, for signaling action of nearby glial cells and other neuronal cells (Farina, Aloisi, and Meinl 2007, Hohlfeld, Kerschensteiner, and Meinl 2007). Cytokines released into the bloodstream also contribute to the increased permeability of the BBB, leading to the recruitment of immune cells from the circulation into the CNS parenchyma. These elements collectively enhance both the initial innate immune responses and subsequent adaptive immunological responses. This cascade has the potential to eliminate infectious or harmful insults and facilitate the restoration of tissue integrity, including scar formation (Engelhardt 2008). In AD and multiple sclerosis (MS), astrocytes play an essential role (Farina, Aloisi, and Meinl 2007, Schwab and McGeer 2008). Astrocytes go through a sequence of structural and functional changes known as astrogliosis when they are exposed to pathogenic circumstances. In astrocyte stimulation, amyloid-peptide (A-) plays a significant function. A distinctive feature of AD involves the accumulation of astrocytes surrounding senile plaques and neurofibrillary tangles (Schwab & McGeer, 2008). When stimulated, astrocytes release various chemokines, attracting microglia and monocyte/macrophages to the plaques. Additionally, they release neurotoxins and pro-inflammatory cytokines, collectively contributing to neurodegeneration (Schwab & McGeer, 2008). In the context of MS, astrocytes play a role by releasing pro-inflammatory cytokines, matrix metalloproteinases (MMP), and free radicals, leading to autoimmune cell recruitment, oligodendrocyte death, and demyelination (Farina et al., 2007). Despite modest expression levels of major histocompatibility complex (MHC) and co-stimulatory molecules even after in vitro stimulation, astrocytes are more inclined to induce T helper 2 (Th2) cell responses rather than T helper 1 (Th1) cell responses. Both microglia and astrocytes produce chemokines that differentially regulate the recruitment of Th1 and Th2 cells. Furthermore, astrocytes have been shown to induce anergy in Th cells through the anti-inflammatory cytokine transforming growth factor (Trajkovic et al., 2004) and the chemokine CXCL12 (Meiron et al., 2008). Alternatively, they may promote the recruitment and proliferation of regulatory T cells (Treg) via the anti-inflammatory cytokine transforming growth factor (Trajkovic et al., 2004). Other anti-inflammatory cytokines or soluble substances secreted by astrocytes, such as IL10, interferons, and neurotrophins, can diminish microglia and T cell activation (Farina, Aloisi, and Meinl 2007, Hohlfeld, Kerschensteiner, and Meinl 2007). The intricate network of interactions involving astrocytes, neurons, microglia, and T cells establishes a delicate balance between pro-inflammatory and anti-inflammatory signals. This modulation influences the outcomes of immune responses within the tissue microenvironment of the CNS. Microglia along with astrocytes play essential role in inflammatory response outcomes and disorders of CNS. These cells not only can boost immune responses and accelerate neurodegeneration, but also can protect brain from inflammation and reduce brain damage.

3.3 NEURAL AND HORMONAL IMPACTS ON THE IMMUNE SYSTEM

3.3.1 Lymphoid Organ Innervation: Sympathetic Versus Parasympathetic Innervation

Limited anatomical and functional evidence exists regarding the involvement of the autonomic nervous system (ANS) in immune modulation (Madden & Felten, 1995). Sympathetic nerves accessing lymphoid organs travel alongside the vasculature but diverge from it, forming "neuroeffector junctions" near parenchymal T lymphocytes and plasma cells (Felten et al., 1985). Neuropeptides, including vasoactive intestinal peptide, substance P, neuropeptide Y, Met-enkephalin, and neurotensin, are either colocalized or not with norepinephrine in the nerve fibers innervating lymphoid organs. While the sympathetic and parasympathetic nervous systems innervate most peripheral organs, the immune system appears to be an exception. With the possible exception of lymphoid tissue in the alimentary and respiratory tract, there is no neuroanatomical evidence for parasympathetic or vagal nerve supply to immunological organs (Nance & Sanders, 2007). Revisiting the concept of lymphoid organs receiving parasympathetic innervation, it was found that cholinergic innervation is present only in the intestinal lamina propria of the gastrointestinal system, originating from enteric neurons. Cholinergic fibers closely interact with macrophages, plasma cells, and lymphocytes in the intestinal mucosa, as well as immune cells in the gut-associated lymphoid tissue, to a lesser extent. Cholinergic innervation is supplied to the spleen by a restricted subset of cholinergic sympathetic postganglionic neurons located in the paravertebral and/or prevertebral chains. Periarteriolar catecholaminergic fibres are frequently connected with these fibres. A few of them come to an end in sections of the white pulp that contain lymphocytes. The spleen contains nonneuronal choline acetyltransferase cells, which are mostly B cells with a few T cells. An additional factor contributing to uncertainty in pinpointing cholinergic innervation of lymphoid organs is the existence of a switch from catecholaminergic to cholinergic transmission in certain sympathetic fibers. This switch is influenced by signals generated by the target being innervated, such as leukaemia inhibitory factor. Although this phenomenon is minor, it has been documented in sympathetic fibers that innervate sweat glands in mammalian footpads (Ernsberger & Rohrer, 1999) and the periosteum (Asmus et al., 2000). Tracey and colleagues has discovered unexpected pharmacological features of anti-inflammatory medication tetravalent guanylhydrazone (CNI-1493), which is used to treat sepsis. CNI-1493 stimulates a neuronal route that connects the brain to peripheral macrophages, according to this discovery. The vagus nerves were alleged to be this neurological pathway since cutting the vagus nerves stopped CNI-1493's anti-inflammatory action while stimulating the peripheral ends of the vagus nerve (Borovikova et al. 2000). Due to the absence of synaptic connections between vagal preganglionic neurons and sympathetic postganglionic neurons, as well as the lack of recruitment of splenic neurons during vagal nerve stimulation, there has been a debate about the existence of indirect vagal innervation of the spleen by splenic nerves (Bratton et al., 2012). This control is aided by interactions between locally generated cytokines and sympathetic nerve fibres (Rogausch et al. 1997, Rogausch et al. 2003). There is a substantial increase in splenic blood flow during sepsis, which promotes the inflammatory reaction, as Tracey discovered. This causes sympathetic nerve fibers to release less norepinephrine and potentially other neuromodulators in response to locally generated IL1(Rogausch et al. 1997). The lymphoid organs are primarily sympathetically innervated, as stated before in this section. The gut lymphoid tissue appears to be the only lymphoid organ that receives parasympathetic innervation. Further investigation is required to explore the potential impact of innervation on immune cell functioning by modifying the blood supply to lymphoid organs.

3.3.2 Catecholamine Effects on Immune Cells

Innate immune cells express the 2, 1, and 2 subtypes of adrenergic receptors, whereas adaptive immune cells predominantly express the 2 subtype (Nance & Sanders, 2007). Activation of 2-adrenergic receptors triggers the formation of cAMP, which subsequently activates either protein kinase A (PKA) or guanine exchange proteins directly activated by cAMP (EPACs) (Kolmus et al., 2015). Chemical sympathetic denervation has been observed to reduce the severity of delayed-type hypersensitivity in vivo, indicating a favorable role of catecholamines in Th1 cell-driven immunological responses (Madden et al., 1989). The responses of macrophages and monocytes to adrenergic stimulation are intricate due to their possession of 1-, 2-, and 2-adrenergic receptors. Catecholamines, synthesized locally, can serve various functions and act as coordinating molecules between physiological and immunological processes when released from autonomic neurons. Normal T cells are capable of producing and releasing epinephrine and norepinephrine, which act in an autocrine or paracrine manner through traditional adrenergic receptors on T cells. Lymphocytes possess the ability to generate and break down catecholamines, releasing additional catecholamines when stimulated (Marino & Cosentino, 2013; Qiu et al., 2004). Catecholamines exert influence on lymphocyte trafficking, cytokine production, local vascular perfusion, and functional activity of lymphocytes, especially in lymphocytes from the mesenteric lymph nodes (Qiu et al., 2004). Upon stimulation, phagocytic cells such as macrophages and neutrophils can release substantial amounts of adrenaline and norepinephrine, linked to increased expression of enzymes involved in catecholamine synthesis and degradation. Experiments demonstrating that depleting T cells and administering reserpine for sympathetic denervation had no effect on the impacts of 2-adrenergic drugs ruled out the possibility that T cells or sympathetic innervation were responsible (Dantzer R., 2018). Moreover, inhibiting tyrosine hydroxylase or dopamine hydroxylase was found to reduce inflammation, while inhibiting the catecholamine-O-methyl transferase enzyme exacerbated lung injury.

3.3.3 Glucocorticoids

3.3.3.1 Anti and pro-inflammatory activities

Glucocorticoids have traditionally been considered as the hormonal mediators responsible for the immunosuppressive effects of stress. These hormones bind to glucocorticoid receptors (GRs) and exert their actions on biological targets. Typically located in the cytosol within a protein complex comprising heat-shock proteins and the FK506 binding protein 52 (p52), GRs undergo dissociation from this complex upon interaction with their ligands. The resulting monomers or dimers then translocate to the nucleus, where they induce anti-inflammatory effects by interacting with nuclear factor kappa-light-chain-enhancer of activated B cells (NF-kB), activator protein (AP)-1, and interferon-regulating factor-3 (IRF-3), ultimately leading to the down-regulation of genes encoding inflammatory mediators, enzymes, and adhesion molecules (Vandevyver et al., 2013). Under specific conditions, glucocorticoids may exhibit pro-inflammatory actions. The anti-inflammatory effects could stem from the suppression of anti-inflammatory factor production and release, or conversely, an increase in the production and release of proinflammatory mediators (Calandra et al., 1995). Exposure to stressors, whether acute or chronic, 24 hours prior to Lipopolysaccharide (LPS) treatment, has been observed to sensitize or prime the inflammatory response to LPS. This priming effect has been noted not only in peripheral macrophages but also in microglia isolated from the hippocampus of stressed rats (Frank et al., 2007). Furthermore, glucocorticoids have the capacity to repress the expression of adhesion molecules, affecting the migration, adhesion, and extravasation of neutrophils at the site of inflammation (Cruz-Topete & Cidlowski, 2015).

3.3.3.2 Autocrine/paracrine signaling

Glucocorticoids, classic hormonal substances originating from the adrenal cortex and distributed into the bloodstream to affect distant target organs, were initially believed to function solely in an endocrine manner. The unexpected discovery of extra-adrenal glucocorticoid synthesis and release, serving as paracrine or autocrine communication signals, was particularly surprising (Talaber et al., 2013). The thymus emerged as the first organ where extra-adrenal steroidogenesis was identified. In mice, glucocorticoids induce significant death of CD4+ CD8+ thymocytes and lead to thymic involution (Gruber et al., 1994). Thymic epithelial cells in mice were found to possess all the necessary components for local corticosterone synthesis and release, as demonstrated in vitro (Pazirandeh et al., 1999). Thymic-derived glucocorticoids play a crucial role in regulating thymocyte development, production of naive T-cells, and the seeding of recent thymic emigrants in the periphery. This function of thymic glucocorticoids in controlling robust and systemic immune responses complements the role of adrenal glucocorticoids (Talaber et al., 2013).

3.3.4 Neuropeptides and other hormones

Neuropeptides, such as enkephalins, a pentapeptide, play a role in regulating the body's neoceptin response. Enkephalins are also referred to as auto ligands because they originate within the system and bind to the body's own opioid receptors. Acting as fine-tuners of the immune response, neuropeptides enhance the activity of immunocytes while suppressing the activities of activated immune cells (Eriksson et al., 2001). In the context of inflammation-mediated pain, there is substantial evidence of immune-to-nerve communication involving immune-derived peptides. Immunocytes containing opioids travel to the site of inflammation, releasing -endorphin, which activates peripheral opioid receptors on sensory nerves, thereby reducing pain. Various signal molecules, including norepinephrine, corticotrophin-releasing hormone, and

IL-1, are implicated in the release of opioid peptides, with their production and release being heightened at the site of inflammation (Stein et al., 2009). The understanding that prolactin and growth hormone influence immunity through a paracrine/autocrine mechanism rather than a hormonal one was established early in the investigation of these peptides' immunological properties. Interestingly, the immunosuppressive drug cyclosporine was found to elevate prolactin levels in the blood, and the dopamine agonist bromocriptine reduced the immunosuppressive characteristics of cyclosporine when administered at doses preventing cyclosporine-induced prolactin increases (Cardon et al., 1984). It is now well-established that signaling cues from immune cells regulate the production and release of hormones from the pituitary gland and lymphocytes. The diverse effects of neuropeptides on immune cells can be attributed to post-transcriptional and post-translational modifications leading to the generation of different isoforms of peptides.

3.4 IMMUNE EFFECTS ON THE NERVOUS SYSTEM

3.4.1 Historical Prospective

Until 1980, there was little known about immunological influence on the neurological system. This is very much besieged by the chemical nature of immunological communication signals such as monokines and interleukins. Monokines, IL1 was the first cytokine was cloned and synthesized in 1980 using recombinant technology. Two separate lines of inquiry led to the discovery of cytokine effects in the brain. The initial field of study was and continues to be fever physiology, an integral part of mainstream physiology. Immunophysiology emerged as the second discipline, taking more time to develop due to its relatively small scope at the intersection of immunology and physiology. In the realm of immunophysiology, researchers found that immunological signals could influence the immune system through a downregulating neuroendocrine feedback loop, which involves cytokine activity in the brain. Similar to any other physiological system in the body, the immune system requires dual levels of regulation: an internal regulatory loop employing the CNS and its communication pathways with the ANS and the neuroendocrine system, as well as an external regulatory loop utilizing the CNS and its communication channels with the rest of the body. Besedovsky and Sorkin proposed that if extrinsic regulation exists, it should involve immunological factors signaling the brain to engage immunomodulatory neuroendocrine factors. Their research was significantly influenced by the discovery that an antibody response was associated with increased activity of the HPA axis and the sympathetic nervous system (SNS). Prior to this finding, the regulation of the immune response was believed to occur solely within the immune system. In experiments with mice and rats injected with sheep red blood cells (a T-cell-dependent antigen), the peak of the immunological response coincided with an elevation in glucocorticoid blood levels (Besedovsky et al., 1975). Furthermore, decreased norepinephrine in the spleen preceded this peak, which was inversely associated with the amplitude of the immunological response (del Rey, Besedovsky, and Sorkin 1984). Besedovsky and Sorkin illustrated the vital role of the glucocorticoid response in antigenic stimulation for functional antigenic competition. Antigenic competition occurs when an immune reaction to a first antigen inhibits the immunological response to a second non-cross-reactive antigen. Animals deficient in adrenaline, incapable of initiating a glucocorticoid response to the initial antigen, do not exhibit this form of competition. In their research, they injected supernatants of in vitro activated lymphocytes into naive mice to identify immunological components signaling the ongoing immune response to the brain neuroendocrine circuits. In this investigation, they identified a "glucocorticoid boosting factor" (Besedovsky *et al.*, 1979), demonstrating that this cytokine activated the pituitary-adrenal axis in a manner similar to the glucocorticoid-enhancing factor. The discovery of IL1's impact on stimulating glucocorticoid release raised questions about its exact site of action. It was crucial to establish that this influence was mediated in the hypothalamus rather than the pituitary or adrenal cortex, a point validated in subsequent research (Berkenbosch *et al.*, 1987). The function of IL1 in a brain-mediated immunoregulatory loop was eventually established, paving the way for more research into the mechanics of these regulatory processes. The rationale behind uncovering the impact of immunological communication signals on the CNS paralleled the reasoning that elucidated the influence of neuroendocrine communication signals on the immune system. Initially, research predominantly focused on long-range interactions between the immune system and the brain, primarily through hormone-like factors. However, there has been a notable shift towards investigating short-range interactions, emphasizing immune factors locally produced within the neurological system, which has progressively garnered more attention.

3.4.2 Cytokine Pyrogenic Effects

The fundamental understanding of fever in viral infection is that it is mediated through endogenous pyrogens released into the systemic circulation by activated mononuclear phagocytes. This explains why most fever experiments used naive experimental animals who were given intravenous injections of endogenous pyrogens. Even though cytokines do not operate as circulating communication agents, this method of injection was nevertheless used when recombinant cytokines became available. The cascade that causes innate immune cells to produce inflammatory cytokines in response to pathogenic bacteria has been discovered. TLRs are specialized monocyte/macrophage membrane receptors that identify specific pathogen molecular moieties known as pathogen-associated molecular patterns (PAMPs) (Ferrandon, Imler, and Hoffmann 2004). PAMPs and danger-associated molecular patterns (DAMPs) originating from injured host cells activate TLRs. This activation initiates a signaling cascade that ultimately results in the synthesis and release of proinflammatory cytokines (Li, D., & Wu, M., 2021). In terms of communication, the consensus view was that proinflammatory cytokines were released into the general circulation by innate immune cells and functioned at a distance. Proinflammatory cytokines could not pass the BBB due to their large molecular size and hydrophilic nature. Instead, they targeted the circumventricular organs, which are areas of the brain that lack a BBB. The organum vasculum laminae terminalis (OVLT) was the target circumventricular organ during the fever reaction. Cytokines stimulated the production of prostaglandins E2 (PGE2) in the OVLT (Blatteis 1990).

3.4.3 Cytokine and HPA

Tracking the neuroendocrine response to an infectious pathogen doesn't offer the same temporal precision as monitoring fluctuations in body temperature. This discrepancy contributes to the frequent characterization of the HPA axis as monophasic rather than multiphasic in its reaction to peripheral immunological stimulation. Nevertheless, it is possible to delineate two distinct phases: an early phase manifesting 1–2 hours after the injection of LPS and a more prolonged response still observable 6 hours later (Elander et al., 2009). Interleukins IL1 and IL2 can operate directly on pituitary cells, causing them to release more pituitary peptides like Adrenocorticotropic hormone (ACTH). The anterior pituitary's follicular-stellate cells are a major source of IL6, which has a paracrine effect on the anterior lobe's hormone production (Denef 2008) and induces pituitary cell adenoma (Sapochnik, Fuertes, and Arzt 2017). Interferon injections can develop glucocorticoid resistance that, in turn, can restrain the activating effect on the HPA axis. Rather than the pituitary gland, glucocorticoid resistance begins in the hippocampus (Gisslinger et al. 1993). However, this is not an artefact of exogenous cytokine (IFNS) administration; the same effect also has been observed during virus infection. Chronic exposure to inflammatory cytokines may lead to glucocorticoid resistance by impairing GR function, involving disruptions in translocation, DNA binding, and altered phosphorylation (Pace, Hu, and Miller 2007).

3.4.4 Cytokine and Behavioural Effects

The sickness behaviour that develops in sick people is the most evident face of an infectious agent's response. Microbial infection is characterized by symptoms such as inactivity, reduced responsiveness to external stimuli, fatigue, decreased appetite, and social withdrawal. These behavioral changes are part of a broader array of disease-control strategies that enable organisms to cope with the high metabolic cost associated with mounting an immune response and fever (Hart, 2011). The neural afferents innervating the site of the body experiencing the inflammatory response convey the signals responsible for mediating sickness behavior to the brain. The vagus nerve serves as the primary neural pathway, particularly when LPS is introduced into the peritoneal cavity or when the dorsal vagal complex is reversibly inactivated through local bupivacaine injection (Marvel et al., 2004). Circulating PAMPs engage brain communication pathways while simultaneously activating macrophage-like cells in circumventricular organs. Consequently, pro-inflammatory cytokines are locally synthesized and gradually diffuse into the brain compartment via volume diffusion, with macrophages and microglia actively recruited "en passant" (Konsman et al., 2002). It's been challenging to pinpoint which parts of the brain are involved in illness behaviour. The broad occurrence of microglial activation throughout the brain in reaction to systemic inflammation contributes to the difficulty in pinpointing specific brain regions. Another factor complicating the identification of brain areas responsible for mediating sickness behavior is the variation in these areas depending on the specific behavioral process influenced by inflammation. Administering injections of LPS or proinflammatory cytokines induces systemic inflammation, which correlates with motivational deficits that are most effectively assessed through effort tasks. In these tasks, animals are required to exert sustained effort to obtain an anticipated reward. Regarding neurotransmitters, inflammatory events exert a notable suppressive impact on the activity of neurons expressing orexin. Orexinergic neurons govern motivational arousal and are found in the perifornical area in the posterior lateral hypothalamus region. GABAergic interneurons expressing neurotensin, which detect systemic inflammation indirectly through alterations in leptin, are inhibited by these changes (Grossberg et al., 2011).

3.4.5 Synaptic Plasticity and the Role of Brain Cytokines

Cytokines are generated in response to peripheral immunological stimuli, both within the brain and the body. Despite their lower expression levels compared to their inducible counterparts, these cytokines play a crucial role in synapse physiology and plasticity. Glial TNF consistently regulates the surface expression of alpha-amino-3-hydroxy-5-methyl-4-isoxazole propionic acid (AMPA) receptors on hippocampal neurons. In cultured hippocampal neurons, TNF enhances the presence of surface AMPA receptors on the synapse plasma membrane. On the contrary, when cultures are treated with a soluble version of TNF receptor 1 to inhibit endogenous TNF, it results in the opposite effect (Beattie et al., 2002). In hippocampus slices, similar results were reported. TNF's effect on AMPA receptor trafficking caused fast alterations in synaptic strength at glutamate excitatory synapses in both cases. TNF also induces an increase in the endocytosis of surface γ-aminobutyric acid (GABA) receptors, leading to a reduction in inhibitory synaptic transmission. The adaptive involvement of TNF in synaptic regulation varies depending on the specific brain region. While cortical pyramidal cells and motor neurons respond to TNF similarly to hippocampal neurons, striatal neurons exhibit a different response. In the dorsolateral striatum's GABAergic medium spiny neurons, TNF induces the internalization of AMPA receptors, thereby decreasing corticostriatal synaptic strength (Lewitus et al., 2014). TNF influences both neurons and astrocytes. TNF can regulate glutamate release in astrocytes when it is present in the blood. Contrary to common belief, microglia, rather than neurons, are responsible for the production of constitutive TNF. Microglia represent the resident macrophage population in the CNS. These cells play a crucial role not only as phagocytic cells in the CNS, activated in response to damage or microbial infection, but also in the development and maintenance of normal synaptic function. Microglia, similar to other tissue macrophages, originate from primitive macrophages in the yolk sac during early development, rather than from blood monocytes or meningeal macrophages (Epelman et al., 2014). The activity-dependent removal or pruning of inactive synapses is essential for the formation of mature neuronal networks. The ability of microglia to absorb synaptic material throughout postnatal development in mice demonstrates their importance in this process (Paolicelli et al. 2011). For this, they utilise a variety of signaling channels, including the innate immune system's traditional complement cascade. Besides the complement cascade, microglial cells communicate with neurons and synapses through various signaling channels.

3.5 CHANGES IN IMMUNE-NERVOUS SYSTEM COMMUNICATION: FROM PHYSIOLOGY TO PATHOLOGY

There exists a bidirectional communication between the brain and the neuronal system, where an effect on one system induces changes in the other. Major depressive disorder, representing a brain illness, and cancer, an example of an immune system pathology, are common instances illustrating reciprocal abnormalities in the communication between the immune and neurological systems, leading to pathophysiology. The precise communication mechanisms between the immune system and the brain in the context of psychiatric diseases remain not fully elucidated, with ongoing research primarily operating at the descriptive level. The current focus has shifted from long-range to short-range communication mechanisms, especially involving the sympathetic nervous system in the context of stress and cancer. These examples underscore the crucial concept that pathological processes develop through a complex network of short- and long-range cellular and molecular communication mechanisms.

3.5.1 Major Depressive Disorder (MDD)

There are many kinds of depressions, including major depressive disorders (MDD). MDD sometimes culminates in treatment resistant depression characterized by increased synaptic availability of serotonin and other CNS monoamines. It is believed that depression is most commonly caused by improper functioning of the neurotransmitter system. On the other hand, the monoaminergic theory of depression falls short in explaining all facets of depression and its response to medication. Consequently, the original hypothesis has undergone reformulation. The contemporary neurobiology of depression underscores the significance of neurotrophic factors and synaptic plasticity, with a particular emphasis on the brain-derived neurotrophic factor (BDNF) acting as a vital mediator of the effects of antidepressant medications (Nestler et al., 2002). This brain-centric perspective on sadness leaves limited space for exploring depression-immunity pathways. The early investigations on immunological functioning in depressed people focused on their apparent immunosuppression, which was indicated by decreased lymphocyte proliferative responses to mitogens, decreased natural killer cell activity, and changes in the numbers of various white blood cell types (Herbert and Cohen 1993). Subsequent investigations revealed a more intricate scenario involving the activation of both the monocytic and lymphocytic arms of cell-mediated immunity. This is substantiated by an increase in the numbers of peripheral blood leukocytes, monocytes, and neutrophils, along with elevated levels of inflammatory cytokines and acute-phase proteins in the bloodstream (Maes et al., 1995). In the late 1990s, research involving patients with kidney cancer and metastatic melanoma, who were administered IFN- and IL2 to stimulate the immune system and assist in eliminating tumor cells resistant to chemotherapy or radiation therapy, began shedding light on the mechanisms of inflammation-associated depression. Sickness symptoms, including fatigue, decreased appetite, and sleep disturbances, are experienced by all patients who receive this type of immunotherapy. However, within a few weeks of medication, one-third to one-half of individuals experience depression symptoms (Capuron and Miller 2004). These clinical findings suggest that depression arises in the context of immune-mediated illness, most likely through a mechanism activated by inflammation. The immune-inducible enzyme kynurenine monooxygenase, located in microglia, transforms brain kynurenine into neurotoxic kynurenine metabolites. The N-methyl-D-aspartate (NMDA) receptor agonist Quinolinic acid is one of these hazardous kynurenine metabolites. Microglia release glutamate extrasynaptically when activated in response to immunological activation. The cystine-glutamate exchanger, which controls the manufacture of the intracellular antioxidant glutathione, is responsible for this release (Figure 3.2). Moreover, in inflammatory conditions, astrocytes experience a downregulation of their high-affinity glutamate transporters. This downregulation reduces the efficiency of astrocytes in absorbing glutamate for recycling through the glutamate-glutamine cycle (Gras et al., 2012). As a result, glutamate spills into the extra synaptic region, activating the extra synaptic NMDA receptor in conjunction with quinolinic acid. This explains why ketamine-induced NMDA receptor blockage can reverse the depression-like behaviour produced by LPS in mice (Walker et al. 2013). It's important to highlight that the identification of this glutamatergic pathway does not negate the possibility of serotonin playing a role in inflammation-induced depression. There is evidence suggesting that inflammation activates the serotonin transporter, leading to a decrease in the amount of serotonin available at synapses (Baganz & Blakely, 2013). The impact of immune mediators released by brain immune-like cells on synapse function has been studied extensively. The brain machinery is believed to be engaged by circulating substances such as kynurenine, either through peripheral IDO activation or pathogen-associated molecular patterns originating from the gut or other barrier organs, particularly in the presence of changes in the microbiota (Slyepchenko et al., 2017). These circulating substances do not necessarily have to directly reach the brain to induce depression. They can activate afferent autonomic nerves, influencing the activity of brain regions associated with emotion processing, as demonstrated by neuroimaging studies in humans who reported experiencing sad moods following typhoid vaccination (Harrison et al., 2009). It remains uncertain whether the mechanisms underlying the onset of inflammation-induced depression also account for depression triggered by other factors, such as stress. The query of how these neuroimmune interactions can commence in the brain without evident inflammation remains unresolved. Even relatively mild stressors, like the trier social stress test, where participants are instructed to prepare and deliver an impromptu speech and engage in a challenging mental exercise in front of a nonresponsive, intimidating audience, have been demonstrated to enhance IL6 production (Rohleder et al., 2012). Individuals who have been subjected to prolonged stress have the same elevated levels of circulating IL6. It's unclear if this IL6 comes from immune or non-immune cells. Elevated IL6 levels induced by stress are more pronounced in individuals who have undergone early life stress and those with a low socioeconomic status. Given that these conditions are associated with persistent, low-grade inflammation and the activation of immune-to-brain communication pathways (Nusslock & Miller, 2016), it is highly probable that psychosocial stress-induced IL6 is produced by activated innate immune cells, likely through a β2-adrenergic-dependent mechanism. The defeat phase typically lasts only a few minutes, after which the mice are usually separated to minimize harm. Following several attacks in these tests, the resident mouse is compelled to cohabit with its assailant, but protection from subsequent attacks is provided by a perforated divider dividing the cage into two compartments. Mice subjected to repeated social defeat exhibit social avoidance and reduced mobility in potentially stressful areas when tested with a new male mouse. Although these behavioral changes more closely resemble anxiety than depression, persistent antidepressant treatment, but not acute antidepressant treatment, can rectify these alterations in behavior (Golden et al., 2011). They are also associated with reduced synaptic plasticity, chromatin remodeling, and transcriptional activation, all of which are responsive to antidepressant treatment. Each incident of defeat triggers a high level of activation in the HPA axis and the sympathetic nervous system (Figure 3.2), which is sustained by cohabitation. Socially defeated mice in the general circulation exhibit a higher presence of highly inflammatory and glucocorticoid-insensitive bone marrow-derived myeloid cells.

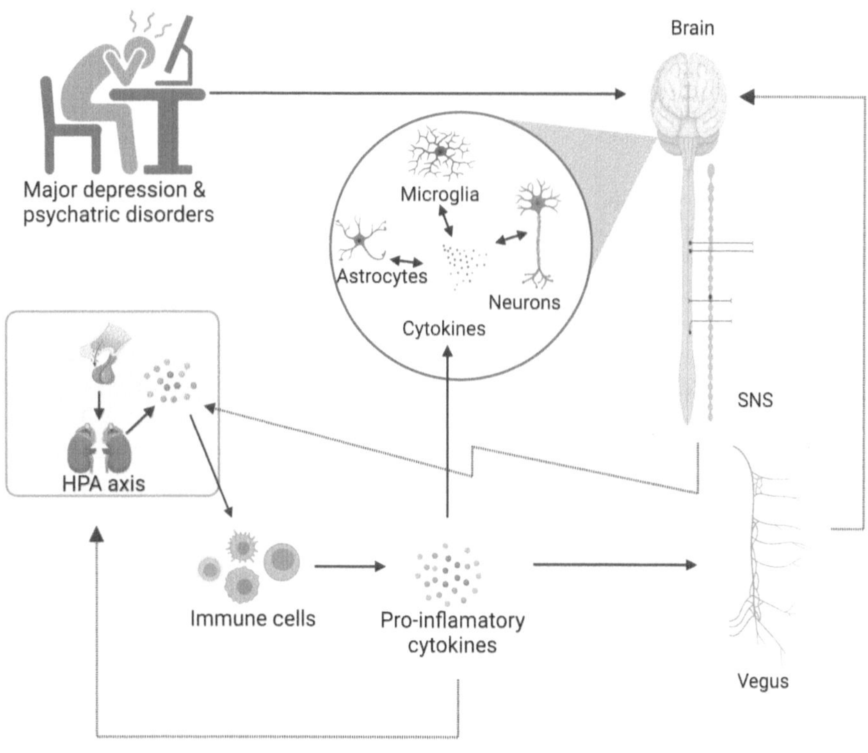

FIGURE 3.2 Major depression and neuro-immune inflammation. Major depression can be caused from any stressful and traumatic events. Cytokines and cheekiness released from CNS can modulate activities in peripheral trough SNS and Vagus. The SNS can modulate HPA axis affecting adrenal cortex, which in turn can modulate immune cell function in a vicious cycle of inflammatory cytokines which in turn can modulate vagus nerve.

The stimulation of β2-adrenergic receptors induces the release of myeloid cells. Chemokines are involved in the recruitment of myeloid cells to the brain. This causes resident microglial cells to become activated, as well as the local generation of inflammatory cytokines that have a deleterious impact on synaptic plasticity (Wohleb et al. 2016). The downstream effects of brain cytokines could also be mediated by microglial cyclooxygenases-1-mediated synthesis of PGE2, acting on prostaglandin E2 receptor 1 (EP1) on GABAergic neurons in the medial prefrontal cortex. EP1 activation would suppress dopaminergic neurons in the mesocortex (Tanaka et al. 2012). Repeated social failure has an impact on immunological memory. The aftermath of repeated social defeat is associated with an increase in memory T-cell populations responding to a viral infection occurring after the stressful episode (Weber et al., 2017). These characteristics extend beyond immunity and encompass immune-to-brain communication as well. The adoptive transfer of lymphocytes from immunocompetent mice that underwent repeated social defeat into stress-naïve, immunocompromised mice has effectively prevented the development of anxiety-like behavior in these mice, likely by mitigating the inflammatory response to defeat stress both peripherally and in the brain (Brachman et al., 2015). These fascinating findings support the hypothesis that psychological stressors might program the adaptive immune system to confer stress tolerance, although the exact pathways are yet unknown.

3.5.2 Cancer

Oncologists collaborating with psychologists have convincingly demonstrated a clear role for the activation of the sympathetic nervous system in modifying the tumor microenvironment and the immune response to the tumor. This was achieved through an elegant series of experimental studies investigating the impact of repeated immobilization stress on the progression and dissemination of cancer in murine models of solid epithelial tumors. The majority of these effects are mediated by β2-adrenergic activation of various local intracellular signaling pathways, including protein kinases, cAMP response element-binding protein (CREB), EPAC, and Mitogen-activated protein kinases (MAPK), which is not surprising. The inhibition of DNA damage repair and p53-associated apoptosis, activation of oncogenic signaling pathways, increased inflammation through elevated expression of proinflammatory cytokines, recruitment of macrophages into the tumor parenchyma, production of tumor growth factors (including transforming growth factor-β, vascular endothelial growth factor, and matrix metalloproteases), activation of mesenchymal cells in the tumor stroma, and promotion of angiogenic factors are all facilitated by sympathetic nervous system signaling. Additionally, sympathetic nervous system signaling contributes to tumor escape from immunosurveillance by blocking type I and II interferon signaling and reducing the cytotoxic action of T lymphocytes and natural killer cells. The idea that the peripheral nervous system might adapt to and contribute to the evolution of a continuing illness process is not entirely new. In conditions like pancreatitis, intricate interactions among neurons, glial cells, immune cells, and endocrine cells have been observed. These interactions regulate inflammation and cancer growth, leading to plastic changes in the ratio of autonomic and sensory fibers innervating the affected organ and the release of various mediators. Pancreatitis and its development into pancreatic cancer have been widely investigated using this type of brain plasticity (Demir, Friess, and Ceyhan 2015). The synthesis of neurotrophic factors by cancer cells, such as nerve growth factor (NGF), drives the development of nerves in solid tumours (Jobling et al. 2015). In vitro experiments involving the knockdown of NGF or its receptors, as well as the administration of an NGF receptor inhibitor, resulted in a reduction of pancreatic cancer cell growth and migration. This dependence on neurotrophins may provide an explanation for the perineural invasion observed in certain solid malignancies, such as head and neck squamous cell carcinoma (Krukowski et al., 2016).

3.6 DISCUSSION

Neuroimmune signaling is a crucial component of the CNS. The neuroimmune factors released from CNS mediate neuroinflammation and modulate mymarid of brain functions, including CNS development, neuronal activity, and endocrine functioning (Frederick, Tavares, and Louveau 2022). Initially, there was a hypothesis suggesting that the immune system functions as a diffuse sensory organ with a specialization in recognizing the non-self, complementing classical organs (Blalock, 1984). However, the current widely accepted understanding is that the nervous system and immune system do not function in isolation. Instead, the CNS mobilizes cellular and molecular components of the immune system. Additionally, the immune system is influenced by neuroendocrine mediators generated by immunocytes. Both the somatosensory and autonomic nervous systems play crucial roles in protecting against infections and injuries through the modulation of immune responses via positive and negative feedback mechanisms. Adverse pathological outcomes may results from failure of one of the interconnected neuroimmunology systems and disruption of cellular and molecular mechanisms regulating the fine-tuned connection between the two systems. Emergence of new small molecule drugs, and designing signaling cue analogue would enable the researcher to understand pathophysiology and to design new therapeutics interventions. In the recent past anti-TNF therapy has been widely explored in the treatment of many pathophysiological anomalies. The TNF antagonist has been explored and established for treating depression among patients with low grade inflammation (Raison et al. 2013) and beta blockers have been used for improvisation and overall survival of patients with ovarian cancer (Watkins et al. 2015). These advancements have clarified the treatment horizon and presented it as a high priority for research advancements for the very benefit of people. There are many aspects of neuro-immune communication still unknown, such as the temporal elements of neuroimmune communications. The mechanistic part of neuronal pain and its psychiatric connections can be elucidated through the understanding of molecular pathways. Neuropathic pain induced by chemotherapeutic agents such as paclitaxel is instigated through activation of innate immune pathways, and its resolution is dependent on an adaptive immune system specifically involving trafficking of $CD8^+$ T cells at the dorsal root ganglia (Krukowski et al. 2016). The immune therapeutics medicine can result in harmful impacts if administered inappropriately in term of dose and time frame of pathophysiology. Imperatively, the neuronal system and immune system are interconnected. The foreseen benefits on one system can be harmful for the other and vice versa. Therefore, the neuron-immune juncture should be understood and taken into consideration prior to initiation of any therapeutic interventions.

APPENDIX-1

TERMINOLOGY

CNS	central nervous system
PRR	pattern recognition receptors
TLR	toll like receptors
TREM	triggering receptors expressed on myeloid cells
HPA axis	hypothalamic-pituitary-adrenal axis
AD	Alzheimer's disease
BBB	blood brain barrier
Treg	regulatory T cells
AP-1	activator protein -1
MS	multiple sclerosis
LPS	lipopolysaccharide
NGF	nerve growth factor
CRF	corticotropin releasing factor
ACTH	adrenocorticotropic hormone
RAGE	receptors for advanced glycation end products
ANS	autonomic nervous system
SNS	sympathetic nervous system
TNF	tumor necrosis factor
MMP	matrix metalloproteinases
MHC	major histocompatibility complex
Th1 cells	T helper 1 cells
Th2 cells	T helper 2 cells
NF-kB	nuclear factor kappa-light-chain-enhancer of activated B cells
IRF-3	interferon-regulating factor-3
PGE2	prostaglandins E2
AMPA	alpha-amino-3-hydroxy-5-methyl-4-isoxazole propionic acid
GABA	γ-aminobutyric acid
MDD	major depressive disorders
NMDA receptors	N-methyl-D-aspartate receptors
cAMP	cyclic adenosine monophosphate
EP1 prostaglandin E2 receptor 1EPAC	exchange protein directly activated by cAMP
CREB	cAMP response element-binding protein
MAPK	mitogen-activated protein kinases
GRs	glucocorticoid receptors

PKA	protein kinase A
DAMPs	danger-associated molecular patterns
PAMPs	pathogen-associated molecular patterns
BDNF	brain-derived neurotrophic factor
OVLT	organum vasculum laminae terminalis

REFERENCES

Ader, R., and N. Cohen. 1975. "Behaviorally conditioned immunosuppression." *Psychosomatic Medicine* 37 (4): 333–40. doi: 10.1097/00006842-197507000-00007

Asmus, S. E., Parsons, S., & Landis, S. C. (2000). Developmental changes in the transmitter properties of sympathetic neurons that innervate the periosteum. *The Journal of neuroscience: the official journal of the Society for Neuroscience*, 20 (4), 1495–1504. https://doi.org/10.1523/JNEUROSCI.20-04-01495.2000

Baganz, N. L., & Blakely, R. D. (2013). A dialogue between the immune system and brain, spoken in the language of serotonin. *ACS chemical neuroscience*, 4 (1), 48–63. https://doi.org/10.1021/cn300186b

Bailey, S. L., P. A. Carpentier, E. J. McMahon, W. S. Begolka, and S. D. Miller. 2006. "Innate and adaptive immune responses of the central nervous system." *Critical Reviews in Immunology* 26 (2): 149–88. doi: 10.1615/critrevimmunol.v26.i2.40

Beattie, E. C., Stellwagen, D., Morishita, W., Bresnahan, J. C., Ha, B. K., Von Zastrow, M., Beattie, M. S., & Malenka, R. C. (2002). Control of synaptic strength by glial TNFalpha. *Science* (New York, N.Y.), 295 (5563), 2282–2285. https://doi.org/10.1126/science.1067859

Becher, B., I. Bechmann, and M. Greter. 2006. "Antigen presentation in autoimmunity and CNS inflammation: How T lymphocytes recognize the brain." *Journal of Molecular Medicine (Berlin, Germany)* 84 (7): 532–43. doi: 10.1007/s00109-006-0065-1

Berkenbosch, F., van Oers, J., del Rey, A., Tilders, F., & Besedovsky, H. (1987). Corticotropin-releasing factor-producing neurons in the rat activated by interleukin-1. *Science* (New York, N.Y.), 238 (4826), 524–526. https://doi.org/10.1126/science.2443979

Besedovsky, H., Sorkin, E., Keller, M., & Müller, J. (1975). Changes in blood hormone levels during the immune response. *Proceedings of the Society for Experimental Biology and Medicine*. Society for Experimental Biology and Medicine (New York, N.Y.), 150 (2), 466–470. https://doi.org/10.3181/00379727-150-39057

Besedovsky, H. O., Del Rey, A., & Sorkin, E. (1979). Antigenic competition between horse and sheep red blood cells as a hormone-dependent phenomenon. *Clinical and experimental immunology*, 37 (1), 106–113.

Bessis, A., C. Bechade, D. Bernard, and A. Roumier. 2007. "Microglial control of neuronal death and synaptic properties." *Glia* 55 (3): 233–8. doi: 10.1002/glia.20459

Black P. H. (2002). Stress and the inflammatory response: a review of neurogenic inflammation. *Brain, behavior, and immunity*, 16 (6), 622–653. https://doi.org/10.1016/s0889-1591(02)00021-1

Blalock, J. E. (1984). The immune system as a sensory organ. *Journal of Immunology*, 132 (3), 1067–1070. https://doi.org/10.4049/jimmunol.132.3.1067

Blatteis, C. M. 1990. "Neuromodulative actions of cytokines." *The Yale Journal of Biology and Medicine* 63 (2): 133–46.

Borovikova, L. V., S. Ivanova, M. Zhang, H. Yang, G. I. Botchkina, L. R. Watkins, H. Wang, N. Abumrad, J. W. Eaton, and K. J. Tracey. 2000. "Vagus nerve stimulation attenuates the systemic inflammatory response to endotoxin." *Nature* 405 (6785): 458–62. doi: 10.1038/35013070

Brachman, R. A., Lehmann, M. L., Maric, D., & Herkenham, M. (2015). Lymphocytes from chronically stressed mice confer antidepressant-like effects to naive mice. *The Journal of neuroscience: the official journal of the Society for Neuroscience*, 35 (4), 1530–1538. https://doi.org/10.1523/JNEUROSCI.2278-14.2015

Bratton, B. O., Martelli, D., McKinley, M. J., Trevaks, D., Anderson, C. R., & McAllen, R. M. (2012). Neural regulation of inflammation: no neural connection from the vagus to splenic sympathetic neurons. *Experimental physiology*, 97 (11), 1180–1185. https://doi.org/10.1113/expphysiol.2011.061531

Calandra, T., Bernhagen, J., Metz, C. N., Spiegel, L. A., Bacher, M., Donnelly, T., Cerami, A., & Bucala, R. (1995). MIF as a glucocorticoid-induced modulator of cytokine production. *Nature*, 377 (6544), 68–71. https://doi.org/10.1038/377068a0

Capuron, L., and A. H. Miller. 2004. "Cytokines and psychopathology: Lessons from interferon-alpha." *Biological Psychiatry* 56 (11): 819–24. doi: 10.1016/j.biopsych.2004.02.009

Cardon, S. B., Larson, D. F., & Russell, D. H. (1984). Rapid elevation of rat serum prolactin concentration by cyclosporine, a novel immunosuppressive drug. *Biochemical and biophysical research communications*, 120 (2), 614–618. https://doi.org/10.1016/0006-291x(84)91299-3

Carpentier, P. A., & Palmer, T. D. (2009). Immune influence on adult neural stem cell regulation and function. *Neuron*, 64 (1), 79–92. https://doi.org/10.1016/j.neuron.2009.08.038

Cruz-Topete, D., & Cidlowski, J. A. (2015). One hormone, two actions: anti- and pro-inflammatory effects of glucocorticoids. *Neuroimmunomodulation*, 22 (1-2), 20–32. https://doi.org/10.1159/000362724

Dantzer R. (2018). Neuroimmune Interactions: From the Brain to the Immune System and Vice Versa. *Physiological reviews*, 98 (1), 477–504. https://doi.org/10.1152/physrev.00039.2016

Demir, I. E., H. Friess, and G. O. Ceyhan. 2015. "Neural plasticity in pancreatitis and pancreatic cancer." *Nature Reviews. Gastroenterology & Hepatology* 12 (11): 649–59. doi: 10.1038/nrgastro.2015.166

Denef, C. 2008. "Paracrinicity: The story of 30 years of cellular pituitary crosstalk." *Journal of Neuroendocrinology* 20 (1): 1–70. doi: 10.1111/j.1365-2826.2007.01616.x

Elander, L., Engström, L., Ruud, J., Mackerlova, L., Jakobsson, P. J., Engblom, D., Nilsberth, C., & Blomqvist, A. (2009). Inducible prostaglandin E2 synthesis interacts in a temporally supplementary sequence with constitutive prostaglandin-synthesizing enzymes in creating the hypothalamic-pituitary-adrenal axis response to immune challenge. *The Journal of neuroscience: the official journal of the Society for Neuroscience*, 29 (5), 1404–1413. https://doi.org/10.1523/JNEUROSCI.5247-08.2009

Engelhardt, B. 2008. "Immune cell entry into the central nervous system: involvement of adhesion molecules and chemokines." *Journal of the Neurological Sciences* 274 (1–2): 23–6. doi: 10.1016/j.jns.2008.05.019

Epelman, S., Lavine, K. J., Beaudin, A. E., Sojka, D. K., Carrero, J. A., Calderon, B., Brija, T., Gautier, E. L., Ivanov, S., Satpathy, A. T., Schilling, J. D., Schwendener, R., Sergin, I., Razani, B., Forsberg, E. C., Yokoyama, W. M., Unanue, E. R., Colonna, M., Randolph, G. J., & Mann, D. L. (2014). Embryonic and adult-derived resident cardiac macrophages are maintained through distinct mechanisms at steady state and during inflammation. *Immunity*, 40 (1), 91–104. https://doi.org/10.1016/j.immuni.2013.11.019

Eriksson, K. S., Sergeeva, O., Brown, R. E., & Haas, H. L. (2001). Orexin/hypocretin excites the histaminergic neurons of the tuberomammillary nucleus. *The Journal of neuroscience: the official journal of the Society for Neuroscience*, 21 (23), 9273–9279. https://doi.org/10.1523/JNEUROSCI.21-23-09273.2001

Ernsberger, U., & Rohrer, H. (1999). Development of the cholinergic neurotransmitter phenotype in postganglionic sympathetic neurons. *Cell and tissue research*, 297 (3), 339–361. https://doi.org/10.1007/s004410051363

Farina, C., F. Aloisi, and E. Meinl. 2007. "Astrocytes are active players in cerebral innate immunity." *Trends in Immunology* 28 (3): 138–45. doi: 10.1016/j.it.2007.01.005

Felten, D. L., Felten, S. Y., Carlson, S. L., Olschowka, J. A., & Livnat, S. (1985). Noradrenergic and peptidergic innervation of lymphoid tissue. *Journal of immunology* (Baltimore, Md.: 1950), 135 (2 Suppl), 755s–765s.

Ferrandon, D., J. L. Imler, and J. A. Hoffmann. 2004. "Sensing infection in Drosophila: Toll and beyond." *Seminars in Immunology* 16 (1): 43–53. doi: 10.1016/j.smim.2003.10.008

Frank, M. G., Baratta, M. V., Sprunger, D. B., Watkins, L. R., & Maier, S. F. (2007). Microglia serve as a neuroimmune substrate for stress-induced potentiation of CNS pro-inflammatory cytokine responses. *Brain, behavior, and immunity*, 21 (1), 47–59. https://doi.org/10.1016/j.bbi.2006.03.005

Frederick, N. M., G. A. Tavares, and A. Louveau. 2022. "Neuroimmune signaling at the brain borders." *Immunological Reviews* 311 (1): 9–25. doi: 10.1111/imr.13126

Gisslinger, H., T. Svoboda, M. Clodi, B. Gilly, H. Ludwig, L. Havelec, and A. Luger. 1993. "Interferon-alpha stimulates the hypothalamic-pituitary-adrenal axis in vivo and in vitro." *Neuroendocrinology* 57 (3): 489–95. doi: 10.1159/000126396

Golden, S. A., Covington, H. E., Berton, O., & Russo, S. J. (2011). A standardized protocol for repeated social defeat stress in mice. *Nature Protocols*, 6 (8), 1183–1191. https://doi.org/10.1038/nprot.2011.361

Gonzalez-Rey, E., Ganea, D., & Delgado, M. (2010). Neuropeptides: keeping the balance between pathogen immunity and immune tolerance. *Current opinion in pharmacology*, 10 (4), 473–481. https://doi.org/10.1016/j.coph.2010.03.003

Gras, G., Samah, B., Hubert, A., Léone, C., Porcheray, F., & Rimaniol, A. C. (2012). EAAT expression by macrophages and microglia: still more questions than answers. *Amino acids*, 42 (1), 221–229. https://doi.org/10.1007/s00726-011-0866-6

Grossberg, A. J., Zhu, X., Leinninger, G. M., Levasseur, P. R., Braun, T. P., Myers, M. G., Jr, & Marks, D. L. (2011). Inflammation-induced lethargy is mediated by suppression of orexin neuron activity. *The Journal of neuroscience: the official journal of the Society for Neuroscience*, 31 (31), 11376–11386. https://doi.org/10.1523/JNEUROSCI.2311-11.2011

Gruber, J., Sgonc, R., Hu, Y. H., Beug, H., & Wick, G. (1994). Thymocyte apoptosis induced by elevated endogenous corticosterone levels. *European journal of immunology*, 24 (5), 1115–1121. https://doi.org/10.1002/eji.1830240516

Guillemin, R. 1978. "Peptides in the brain: The new endocrinology of the neuron." *Science* 202 (4366): 390–402. doi: 10.1126/science.212832

Harrison, N. A., Brydon, L., Walker, C., Gray, M. A., Steptoe, A., & Critchley, H. D. (2009). Inflammation causes mood changes through alterations in subgenual cingulate activity and mesolimbic connectivity. *Biological psychiatry*, 66 (5), 407–414. https://doi.org/10.1016/j.biopsych.2009.03.015

Hart B. L. (2011). Behavioural defences in animals against pathogens and parasites: parallels with the pillars of medicine in humans. *Philosophical transactions of the Royal Society of London. Series B, Biological sciences*, 366 (1583), 3406–3417. https://doi.org/10.1098/rstb.2011.0092

Herbert, T. B., and S. Cohen. 1993. "Depression and immunity: A meta-analytic review." *Psychological Bulletin* 113 (3): 472–86. doi: 10.1037/0033-2909.113.3.472

Hohlfeld, R., M. Kerschensteiner, and E. Meinl. 2007. "Dual role of inflammation in CNS disease." *Neurology* 68 (22 Suppl 3): S58–63; discussion S91-6. doi: 10.1212/01.wnl.0000275234.43506.9b

Jobling, P., J. Pundavela, S. M. Oliveira, S. Roselli, M. M. Walker, and H. Hondermarck. 2015. "Nerve-cancer cell cross-talk: A novel promoter of tumor progression." *Cancer Research* 75 (9): 1777–81. doi: 10.1158/0008-5472.CAN-14-3180

Kolmus, K., Tavernier, J., & Gerlo, S. (2015). β2-Adrenergic receptors in immunity and inflammation: stressing NF-κB. *Brain, behavior, and immunity*, 45, 297–310. https://doi.org/10.1016/j.bbi.2014.10.007

Konsman, J. P., Parnet, P., & Dantzer, R. (2002). Cytokine-induced sickness behaviour: mechanisms and implications. *Trends in neurosciences*, 25 (3), 154–159. https://doi.org/10.1016/s0166-2236(00)02088-9

Krukowski, K., N. Eijkelkamp, G. Laumet, C. E. Hack, Y. Li, P. M. Dougherty, C. J. Heijnen, and A. Kavelaars. 2016. "CD8+ T cells and endogenous IL-10 are required for resolution of chemotherapy-induced neuropathic pain." *The Journal of Neuroscience* 36 (43): 11074–11083. doi: 10.1523/JNEUROSCI.3708-15.2016

Lewitus, G. M., Pribiag, H., Duseja, R., St-Hilaire, M., & Stellwagen, D. (2014). An adaptive role of TNFα in the regulation of striatal synapses. *The Journal of neuroscience: the official journal of the Society for Neuroscience*, 34 (18), 6146–6155. https://doi.org/10.1523/JNEUROSCI.3481-13.2014

Li, D., & Wu, M. (2021). Pattern recognition receptors in health and diseases. *Signal transduction and targeted therapy*, 6 (1), 291. https://doi.org/10.1038/s41392-021-00687-0

Madden, K. S., & Felten, D. L. (1995). Experimental basis for neural-immune interactions. *Physiological reviews*, 75 (1), 77–106. https://doi.org/10.1152/physrev.1995.75.1.77

Madden, K. S., Felten, S. Y., Felten, D. L., Sundaresan, P. R., & Livnat, S. (1989). Sympathetic neural modulation of the immune system. I. Depression of T cell immunity in vivo and vitro following chemical sympathectomy. *Brain, behavior, and immunity*, 3 (1), 72–89. https://doi.org/10.1016/0889-1591(89)90007-x

Maes, M., Meltzer, H. Y., Bosmans, E., Bergmans, R., Vandoolaeghe, E., Ranjan, R., & Desnyder, R. (1995). Increased plasma concentrations of interleukin-6, soluble interleukin-6, soluble interleukin-2 and transferrin receptor in major depression. *Journal of affective disorders*, 34 (4), 301–309. https://doi.org/10.1016/0165-0327(95)00028-l

Marino, F., & Cosentino, M. (2013). Adrenergic modulation of immune cells: an update. *Amino acids*, 45 (1), 55–71. https://doi.org/10.1007/s00726-011-1186-6

Marvel, F. A., Chen, C. C., Badr, N., Gaykema, R. P., & Goehler, L. E. (2004). Reversible inactivation of the dorsal vagal complex blocks lipopolysaccharide-induced social withdrawal and c-Fos expression in central autonomic nuclei. *Brain, behavior, and immunity*, 18 (2), 123–134. https://doi.org/10.1016/j.bbi.2003.09.004

Meiron, M., Zohar, Y., Anunu, R., Wildbaum, G., & Karin, N. (2008). CXCL12 (SDF-1alpha) suppresses ongoing experimental autoimmune encephalomyelitis by selecting antigen-specific regulatory T cells. *The Journal of experimental medicine*, 205 (11), 2643–2655. https://doi.org/10.1084/jem.20080730

Nance, D. M., & Sanders, V. M. (2007). Autonomic innervation and regulation of the immune system (1987-2007). *Brain, behavior, and immunity*, 21 (6), 736–745. https://doi.org/10.1016/j.bbi.2007.03.008

Napoli, I., & Neumann, H. (2009). Microglial clearance function in health and disease. *Neuroscience*, 158 (3), 1030–1038. https://doi.org/10.1016/j.neuroscience.2008.06.046

Nestler, E. J., Barrot, M., DiLeone, R. J., Eisch, A. J., Gold, S. J., & Monteggia, L. M. (2002). Neurobiology of depression. *Neuron*, 34 (1), 13–25. https://doi.org/10.1016/s0896-6273(02)00653-0

"Neuroimmune communication." n.d. *Nature Neuroscience* 20 (2):127. doi: 10.1038/nn.4496

Nusslock, R., & Miller, G. E. (2016). Early-Life Adversity and Physical and Emotional Health Across the Lifespan: A Neuroimmune Network Hypothesis. *Biological psychiatry*, 80 (1), 23–32. https://doi.org/10.1016/j.biopsych.2015.05.017

Pace, T. W., F. Hu, and A. H. Miller. 2007. "Cytokine-effects on glucocorticoid receptor function: Relevance to glucocorticoid resistance and the pathophysiology and treatment of major depression." *Brain, Behavior, and Immunity* 21 (1): 9–19. doi: 10.1016/j.bbi.2006.08.009

Paolicelli, R. C., G. Bolasco, F. Pagani, L. Maggi, M. Scianni, P. Panzanelli, M. Giustetto, T. A. Ferreira, E. Guiducci, L. Dumas, D. Ragozzino, and C. T. Gross. 2011. “Synaptic pruning by microglia is necessary for normal brain development.” *Science* 333 (6048): 1456–8. doi: 10.1126/science.1202529

Pazirandeh, A., Xue, Y., Rafter, I., Sjövall, J., Jondal, M., & Okret, S. (1999). Paracrine glucocorticoid activity produced by mouse thymic epithelial cells. *FASEB journal: official publication of the Federation of American Societies for Experimental Biology*, 13 (8), 893–901. https://doi.org/10.1096/fasebj.13.8.893

Perry, V. H. 2004. “The influence of systemic inflammation on inflammation in the brain: Implications for chronic neurodegenerative disease.” *Brain, Behavior, and Immunity* 18 (5): 407–13. doi: 10.1016/j.bbi.2004.01.004

Qiu, Y. H., Peng, Y. P., Jiang, J. M., & Wang, J. J. (2004). Expression of tyrosine hydroxylase in lymphocytes and effect of endogenous catecholamines on lymphocyte function. *Neuroimmunomodulation*, 11 (2), 75–83. https://doi.org/10.1159/000075316

Raison, C. L., R. E. Rutherford, B. J. Woolwine, C. Shuo, P. Schettler, D. F. Drake, E. Haroon, and A. H. Miller. 2013. “A randomized controlled trial of the tumor necrosis factor antagonist infliximab for treatment-resistant depression: The role of baseline inflammatory biomarkers.” *JAMA Psychiatry* 70 (1): 31–41. doi: 10.1001/2013.jamapsychiatry.4

del Rey, A., H. Besedovsky, and E. Sorkin. 1984. “Endogenous blood levels of corticosterone control the immunologic cell mass and B cell activity in mice.” *Journal of Immunology* 133 (2): 572–5.

Rivest S. (2009). Regulation of innate immune responses in the brain. *Nature reviews. Immunology*, 9 (6), 429–439. https://doi.org/10.1038/nri2565

Rogausch, H., A. del Rey, A. Kabiersch, W. Reschke, J. Ortel, and H. Besedovsky. 1997. “Endotoxin impedes vasoconstriction in the spleen: Role of endogenous interleukin-1 and sympathetic innervation.” *The American Journal of Physiology* 272 (6 Pt 2): R2048–54. doi: 10.1152/ajpregu.1997.272.6.R2048

Rogausch, H., D. Zwingmann, M. Trudewind, A. del Rey, K. H. Voigt, and H. Besedovsky. 2003. “Local and systemic autonomic nervous effects on cell migration to the spleen.” *Journal of Applied Physiology (Bethesda, MD: 1985)* 94 (2): 469–75. doi: 10.1152/japplphysiol.00411.2002

Rohleder, N., Aringer, M., & Boentert, M. (2012). Role of interleukin-6 in stress, sleep, and fatigue. *Annals of the New York Academy of Sciences*, 1261, 88–96. https://doi.org/10.1111/j.1749-6632.2012.06634.x

Sapochnik, M., M. Fuertes, and E. Arzt. 2017. “Programmed cell senescence: Role of IL-6 in the pituitary.” *Journal of Molecular Endocrinology* 58 (4): R241–R253. doi: 10.1530/JME-17-0026

Schwab, C., and P. L. McGeer. 2008. “Inflammatory aspects of Alzheimer disease and other neurodegenerative disorders.” *Journal of Alzheimer’s Disease* 13 (4): 359–69. doi: 10.3233/jad-2008-13402

Slyepchenko, A., Maes, M., Jacka, F. N., Köhler, C. A., Barichello, T., McIntyre, R. S., Berk, M., Grande, I., Foster, J. A., Vieta, E., & Carvalho, A. F. (2017). Gut Microbiota, Bacterial Translocation, and Interactions with Diet: Pathophysiological Links between Major Depressive Disorder and Non-Communicable Medical Comorbidities. *Psychotherapy and psychosomatics*, 86 (1), 31–46. https://doi.org/10.1159/000448957

Stein, C., Clark, J. D., Oh, U., Vasko, M. R., Wilcox, G. L., Overland, A. C., Vanderah, T. W., & Spencer, R. H. (2009). Peripheral mechanisms of pain and analgesia. *Brain research reviews*, 60 (1), 90–113. https://doi.org/10.1016/j.brainresrev.2008.12.017

Talabér, G., Jondal, M., & Okret, S. (2013). Extra-adrenal glucocorticoid synthesis: immune regulation and aspects on local organ homeostasis. *Molecular and cellular endocrinology*, 380 (1-2), 89–98. https://doi.org/10.1016/j.mce.2013.05.007

Tambuyzer, B.R., Ponsaerts, P. and Nouwen, E.J. (2009), Microglia: gatekeepers of central nervous system immunology. *Journal of Leukocyte Biology*, 85: 352–370. https://doi.org/10.1189/jlb.0608385

Tanaka, K., T. Furuyashiki, S. Kitaoka, Y. Senzai, Y. Imoto, E. Segi-Nishida, Y. Deguchi, R. M. Breyer, M. D. Breyer, and S. Narumiya. 2012. "Prostaglandin E2-mediated attenuation of mesocortical dopaminergic pathway is critical for susceptibility to repeated social defeat stress in mice." *The Journal of Neuroscience* 32 (12): 4319–29. doi: 10.1523/JNEUROSCI.5952-11.2012

Trajkovic, V., Vuckovic, O., Stosic-Grujicic, S., Miljkovic, D., Popadic, D., Markovic, M., Bumbasirevic, V., Backovic, A., Cvetkovic, I., Harhaji, L., Ramic, Z., & Mostarica Stojkovic, M. (2004). Astrocyte-induced regulatory T cells mitigate CNS autoimmunity. *Glia*, 47 (2), 168–179. https://doi.org/10.1002/glia.20046

Turnbull, A. V., & Rivier, C. L. (1999). Regulation of the hypothalamic-pituitary-adrenal axis by cytokines: actions and mechanisms of action. *Physiological reviews*, 79 (1), 1–71. https://doi.org/10.1152/physrev.1999.79.1.1

Vandevyver, S., Dejager, L., Tuckermann, J., & Libert, C. (2013). New insights into the anti-inflammatory mechanisms of glucocorticoids: an emerging role for glucocorticoid-receptor-mediated transactivation. *Endocrinology*, 154 (3), 993–1007. https://doi.org/10.1210/en.2012-2045

Walker, A. K., D. P. Budac, S. Bisulco, A. W. Lee, R. A. Smith, B. Beenders, K. W. Kelley, and R. Dantzer. 2013. "NMDA receptor blockade by ketamine abrogates lipopolysaccharide-induced depressive-like behavior in C57BL/6J mice." *Neuropsychopharmacology* 38 (9): 1609–16. doi: 10.1038/npp.2013.71

Watkins, J. L., P. H. Thaker, A. M. Nick, L. M. Ramondetta, S. Kumar, D. L. Urbauer, K. Matsuo, K. C. Squires, R. L. Coleman, S. K. Lutgendorf, P. T. Ramirez, and A. K. Sood. 2015. "Clinical impact of selective and nonselective beta-blockers on survival in patients with ovarian cancer." *Cancer* 121 (19): 3444–51. doi: 10.1002/cncr.29392

Weber, M. D., Godbout, J. P., & Sheridan, J. F. (2017). Repeated Social Defeat, Neuroinflammation, and Behavior: Monocytes Carry the Signal. *Neuropsychopharmacology: official publication of the American College of Neuropsychopharmacology*, 42 (1), 46–61. https://doi.org/10.1038/npp.2016.102

Wohleb, E. S., T. Franklin, M. Iwata, and R. S. Duman. 2016. "Integrating neuroimmune systems in the neurobiology of depression." *Nature Reviews. Neuroscience* 17 (8): 497–511. doi: 10.1038/nrn.2016.69

4 Microbiota in Neuroimmune Regulations

Insights into Gut Neuropeptides and Gut Hormones

Ashwin Kumar Jainarayanan
The University of Oxford, Oxford, UK

Mirudula Elanchezhian
Indian Institute of Science Education and Research, Mohali, India

Ranjeet Singh Mahla
The University of Oxford, Oxford, UK

4.1 INTRODUCTION

In recent years, the microbiota-neuro-immune axes have emerged as a captivating focal point in the realm of scientific investigation, drawing the keen interest of researchers and medical practitioners alike. This intricate axis sheds a brilliant light on the complex and interdependent interactions among three integral components: the gut microbiota, the nervous system, and the immune system. Situated within this framework is the gastrointestinal (GI) tract, a dynamic domain brimming with indispensable functions. Among its multifaceted roles, the GI tract masterfully manages the delicate balance of water and electrolytes, orchestrates the intricate dance of food digestion, and efficiently absorbs the vital nutrients that sustain life.

What elevates this environment to a captivating dimension is its profound synergy with a diverse spectrum of beneficial microorganisms. These microscopic allies wield immense influence, actively participating in the optimization of vital physiological processes. Notably, they facilitate seamless digestion, contribute to the metabolism of foreign substances (known as xenobiotics), and play an active role in

DOI: 10.1201/9781003295471-4

safeguarding the body against the lurking threats of pathobionts—microorganisms with the potential for pathogenicity. In essence, these microorganisms emerge as key players in the preservation of overall health.

The orchestration of the neural, immunological, and microbiota constituents culminates in an intricate defense mechanism of unparalleled sophistication. This mechanism assumes the role of a vigilant guardian, skillfully navigating the multifarious challenges posed by potential pathobionts. Operating within the expansive landscape of the GI tract, this defense system assumes a central position, facilitating not only the assimilation of essential nutrients but also nurturing the broader canvas of digestion and holistic well-being.

Beneath the multifaceted tapestry of the GI tract's functions resides the Enteric Nervous System (ENS), a remarkably intricate network comprising extrinsic nerves, intrinsic neurons, and supportive glial cells. Often referred to as the "second brain," the ENS orchestrates and oversees the seamless functioning of the GI domain. In a perpetual state of vigilant surveillance, the ENS maintains an unwavering commitment to the impeccable operation of the gut, poised to counteract any disruptions that may arise and, in doing so, safeguards the delicate equilibrium of the entire body.

The symphony of interactions among the intricate tapestry of the GI tract's environment, the ENS, and the immune system aptly showcases the complexity inherent in the human body's defense strategies. Their harmonious collaboration assumes a position of paramount significance, constituting the very foundation upon which health and equilibrium rest.

Moreover, recent research has cast a luminous spotlight on the central role of the ENS as a pivotal mediator within the intricate interplay between the gut's microbiota and the immune system (Marklund 2022; Wang et al. 2022). Serving as an interpreter of sorts, the ENS deciphers the subtle cues emanating from the microbiota and deftly translates them into orchestrated immune responses. This newfound understanding illuminates the exceptional aptitude of the ENS in bridging the communication chasm that exists between the microbiota and the immune defenses. This revelation not only contributes to the broader comprehension of gut health and immune responses but also augments the potential for ground-breaking therapeutic interventions.

In essence, the microbiota-neuro-immune axes provide an enlightening window into the elaborate and interconnected mechanisms that underlie the human body's multifaceted responses to diverse challenges. As the journey of exploration within this realm continues to unfold, the prospect of pioneering therapeutic strategies aimed at the preservation of health and the prevention of diseases beckons with profound promise. This axis is not merely a scientific curiosity, but a portal into the intricate symphony that orchestrates the body's intricate defense strategies.

4.2 REGULATION OF MICROBIOTA BY IMMUNE SYSTEM USING BARRIER AND IMMUNE EXCLUSION

4.2.1 Neuro-Immune Sensing at Barrier Surfaces

The human GI system stands as a dynamic ecosystem, hosting a diverse and intricate collection of microbes that significantly impact essential homeostatic mechanisms, ultimately steering the body's responses to physiological shifts. Operating in synergy with the immune and nervous systems, this intricate microbial network forms a remarkable axis responsible for upholding equilibrium. Central to this complex interplay are the immune cells of the innate system, armed with specialized pathogen recognition receptors. These receptors, termed pattern recognition receptors (PRRs), enable these immune cells to discern pathogenic agents within the gut by detecting specific markers called pathogen-associated, molecular patterns (PAMPs). Recognizable molecules within this category encompass lipopolysaccharide (LPS) and CpG. Impressively, these immune cells also sense damage-associated molecular patterns, triggered by tissue damage. This inherent immune vigilance serves as a critical line of defense, rapidly identifying and responding to potential threats within the gut milieu.

Strategically positioned at the junction between the gut and the external environment, certain cells function as frontline sentinels, swiftly detecting and relaying initial disruptions occurring within the epithelial layer. Among these cells are skin Langerhans cells, intestinal CX3CR1 mononuclear phagocytes, and $CD103^+$ epithelial dendritic cells. These sentinel cells play a pivotal role in ensuring that the gut environment remains in a state of dynamic balance (Veiga-Fernandes and Mucida 2016).

Adding another layer of complexity, epithelial cells, often associated with innate immunity, contribute to this intricate interplay. These cells possess a range of immune receptors, including PRRs that identify PAMPs. Beyond their immune functions, these cells forge a neural-immune connection. They act as intermediaries, transmitting immune signals to nearby neurons, thus establishing a link between the immune and nervous systems. Specialized epithelial cells, such as enteroendocrine or enterochromaffin cells (ECs), further enrich the conversation between the gut and immune systems. These cells stand out for their production of a diverse array of molecules, including hormones, neurotransmitters, and peptides. Noteworthy examples encompass Glucagon-related peptides, peptide YY, gastric inhibitory polypeptide, and the bulk of the body's serotonin (Gribble and Reimann 2019). This serotonin wields a profound influence on immune cells residing in both the epithelial and lamina propria compartments.

Notably, the secretion profile of these specialized ECs showcases intriguing adaptability. For instance, the presence of metabolites derived from spore-forming bacteria can stimulate serotonin production in colonic ECs. These findings underscore the dynamic interplay between microbial products and the gut's epithelial components (Yano et al. 2015).

Recent experiments, particularly those involving germ-free mice, have unveiled the remarkable influence of the gut microbiota on enteric neurons. The transplantation of human microbiota into these mice yielded significant effects on the activity of enteric neurons (Dey et al. 2015). Leveraging advanced microscopy techniques, recent scientific research has revealed a tangible interaction between the cytoplasmic extensions of ECs and the neurons that innervate both the small intestine and colon (Bohorquez et al. 2015). This revelation adds depth to our comprehension of the intricate interplay between the gut's microbial community and the neural components pivotal to its functioning.

The interconnection between microbes and the neuronal system is believed to hinge on Toll-like receptors (TLRs), which are expressed by both enteric neurons (TLR2 & TLR4) and glial cells (TLR 2, 3, 4 & 7). These receptors serve as direct detectors of microbial products, sparking a cascade of interactions. Concurrently, microbial products exert an influence on the activities of immune cells, subsequently shaping the development and operation of neuroglia (Brown et al. 2019). A compelling illustration of this interaction lies in the activation of muscularis macrophages (MM) by microbiota, triggering the production of bone morphogenetic protein 2 (BMP-2), a cytokine that regulates enteric neuron activity (Muller et al. 2014). Furthermore, bacterial infections are observed to precede the activation of sympathetic ganglia, steering macrophages towards a functional profile that safeguards tissues.

The comprehensive role of the gut microbiota in shaping intestinal function and maintaining equilibrium emerges distinctly. It orchestrates the maturation and intricate interplay between the gut's nervous and immune systems. The convergence of microbial products, immune cells, and neural elements underscores the substantial influence the gut microbiota exerts over the broader functioning of the body.

4.3 MICROBIOTA IN DEVELOPMENT AND REGULATION OF NEURO-IMMUNE SYSTEM

4.3.1 Microbiota and Nervous System

Within the complex fabric of the Central Nervous System (CNS), an extraordinary ecosystem thrives—a community of innate immune cells primarily constituted by microglia cells, strategically positioned at the heart of the CNS parenchyma. These microglia cells, originating from precursors nestled within the bone marrow, undertake a migratory odyssey to the CNS during the embryonic phase. Symbolic of a foundational stronghold, microglia cells shoulder the crucial duty of addressing and absorbing disturbances in CNS homeostasis, all while maintaining a state of tranquil balance.

Their role stretches beyond a singular aspect, encompassing a wide range of functions that include vigilant immune surveillance, the reinforcement of resilient defense mechanisms, and the intricate orchestration of CNS equilibrium. In the developmental narrative of the CNS, microglia cells emerge as central figures, executing essential tasks like the phagocytosis of residual materials left behind by apoptotic cells and actively shaping the framework of synaptic connections (Shigemoto-Mogami and Sato 2021). This role gains further prominence through their expression of Major Histocompatibility Complex (MHC) molecules and the discharge of proinflammatory cytokines, providing them with a distinct capacity to engage in complex intercellular dialogues with other immune cells.

However, the saga of these microglia cells unfolds with even more captivating layers. The enigma surrounding their maturation and activation amid homeostatic conditions was unraveled through the pioneering endeavors of Enry *et al.* This seminal exploration illuminated the considerable influence of the host microbiota in this intricate process. A revelation of profound import, it underscores the host microbiota's transcendence of its conventional role, emerging as a significant conductor in the postnatal maturation of the endocrine system—a foundational cornerstone upon which the functionality of the CNS is firmly established (Erny et al. 2015).

Study with germ-free mice where elevated stress responses were observed, serves as an initial good example showing influence of microbiota (Diaz Heijtz et al. 2011). The intestinal microbiota affect central levels of brain-derived neurotropic factor and behavior in mice. These germ-free mice have several changes in adulthood pertaining to plasticity related proteins, perturbation in neurotransmitter levels, anxiety and social behaviour (Stilling, Dinan, and Cryan 2014). Moreover, it has also been shown that the host microbiota regulates and influences the permeability of blood brain barrier (BBB) (Braniste et al. 2014) along with that is crucial for adult hippocampal neurogenesis (Ogbonnaya et al. 2015).

From a comprehensive standpoint, the involvement of the host microbiota in shaping CNS function and development introduces a fresh perspective on brain health. The interplay between gut microbiota and the CNS weaves an intricate network of effects that stretch beyond traditional limits, emphasizing the significant influence of microbial communities on our physical and cognitive health. As we explore these complexities more deeply, we uncover novel paths for potential interventions focused on enhancing brain health by adjusting the gut microbiota. This discovery not just enhances our understanding of neuro-immune connections but also offers potential for inventive therapeutic approaches to enhance brain health and resilience.

4.4 INTERACTION OF MICROBIOTA WITH NEURO-IMMUNE SYSTEM

The gut microbiota significantly influences the maturation and progression of the immune system, assuming a pivotal role as an indispensable guide. Fascinatingly, investigations involving germ-free neonatal mice have unveiled a reduction in the expression of genes intricately associated with immune response co-stimulation. This deficiency coincides with disruptions in fundamental immunological processes, including compromised phagocytosis and the function of cytotoxic T-killer cells. Furthermore, these mice exhibit imbalances in the interplay between dendritic cells and T lymphocytes, encompassing the intricate modulation of anti-inflammatory regulatory T cells (Tregs) (Sanidad and Zeng 2020).

This intricate process revolves around microbial metabolites, particularly short-chain fatty acids (SCFAs) such as acetic acid, propionic acid, and butyric acid, generated through the breakdown of dietary fibers within the host's gut. Recent studies have unveiled the profound impact of SCFAs on the immune system. Notably, in conditions like inflammatory bowel disease (IBD), mice display diminished SCFA levels, coupled with heightened intestinal inflammation (Parada Venegas et al. 2019). Moreover, the absence of G protein-coupled receptor, GPR43, a specialized sensor attuned to detecting SCFAs, exacerbates intestinal disorders in mice, a fact underscored by the seminal research of Maslowski et al. (2009).

Further insights emerge from studies involving mice fed high-fiber diets, resulting in elevated SCFA levels. These mice exhibit an enhanced defense mechanism against allergic lung inflammation, attributed to propionate's influence on the hematopoiesis of innate immune cells, a key insight brought to light by Trompette et al. (2014). SCFAs wield a direct influence on the development and functionality of Tregs, critical in curbing uncontrolled inflammation. Remarkably, germ-free mice exposed to SCFAs, especially propionate, undergo an augmentation in the quantity, proportion, and enhanced efficiency of Tregs, a discovery illuminated by Smith et al. (2013).

In culmination, these collective explorations underscore the crucial role of gut microbial metabolites like SCFAs in fortifying both the innate and adaptive immune systems. The intricate interplay between these metabolites and immune components unveils a novel dimension in comprehending the gut-immune axis, illuminating the complex interdependencies steering the body's intricate immune responses. This newfound knowledge opens avenues for innovative strategies in modulating immune function and underscores the profound interconnectedness of the gut microbiota with broader physiological systems.

4.5 INTERACTION OF MICROBIOTA WITH NEURO-IMMUNE SYSTEM

The groundbreaking investigation conducted by Erny et al. (Erny et al. 2015) has elegantly exposed the intricate role played by the gut microbiota in influencing microglia activation in mice. By employing a sophisticated transcriptomics approach, the researchers adeptly identified a cluster of downregulated gene expressions within microglia from germ-free mice, with a substantial portion of these genes intimately tied to mechanisms of cellular activation.

Noteworthy among these genes are:

- MAPK8 (JNK1): Demonstrably crucial for tumor necrosis factor (TNF)-alpha induced apoptosis, with a speculated role in critical processes like T cell proliferation, apoptosis, and differentiation.
- FCGR2B: Encoding a low-affinity receptor for immunoglobulin gamma complexes, pivotal in the phagocytosis of immune complexes and the regulation of antibody production by B-cells. Genetic variations in this gene have been associated with an elevated susceptibility to systemic lupus erythematosus.
- IL1a: Serving as a pivotal driver of inflammation and also contributing to the promotion of fever and the initiation of sepsis.
- LY86: Intriguingly hypothesized to orchestrate the innate immune response to bacterial LPS, potentially in cooperation with CD180 and TLR4, consequently inducing cytokine production.
- CD86: Present on multiple antigen-presenting cells such as macrophages and B cells. Working in conjunction with CD80, CD86 provides essential costimulatory signals that play a pivotal role in T cell activation and survival. It's noteworthy that Cd86 can mediate diverse outcomes, including self regulation and cell-cell association, or attenuation of regulation and cell-cell disassociation, contingent upon the context.

- HIF1A: Significantly implicated as a mastermind orchestrating the transcriptional responses associated with cellular and developmental reactions to hypoxia. Changes in HIF1A expression have been intricately connected to cancer development, as well as various pathophysiologies, cell survival, and the invasive properties of tumors.

Furthermore, the study observed a notable increase in the population of microglia, accompanied by a surge in the number of segments, branching patterns, and terminal points within the brains of germ-free mice. Simultaneously, gene transcription analysis illuminated a unique response of microglia to bacterial stimuli, evident both in germ-free mice and mature mice exposed to a broad-spectrum antibiotic cocktail. Remarkably, reintroducing the microbiota from a control-specific pathogen-free mouse into these mice rectified the anomalies in the microglia phenotype, highlighting the profound influence of the microbiota on the functional behavior and structure of microglia.

Diversifying the study's focus, it delved into the pivotal role of SCFAs, bacterial metabolites renowned for their substantial impact on metabolism, immunity, and brain health. Intriguingly, these SCFAs displayed the capacity to address the deficits in BBB signaling observed within the germ-free mouse model. Impressively, the deficiencies in microglia density, morphology, and maturity were effectively reversed through the administration of a mixture of SCFAs (comprising sodium propionate, sodium butyrate, and sodium acetate) via the drinking water of germ-free mice (Dalile et al. 2019). The precise mechanism through which SCFAs operate in this context remains a tantalizing question, particularly given the absence of specific receptors for SCFAs within the brain.

Expanding the horizons of the research, an exploration into the direct interaction between the gut microbiota and the ENS was pursued. This intricate system comprises cells vital for creating diffusion barriers, including astroglia, between intestinal capillaries and ganglia. The microorganisms within the gut engage in a multifaceted defense strategy, involving competition for colonization sites against potential pathogens and proactive initiation of the immune system. A pivotal facet is their role in modulating immune system activity, thereby diminishing the risk of autoimmune disorders. The absence of this inherent immune regulation due to pathogen colonization within the gut has been linked to the emergence of mental disorders, including autism, schizophrenia, depression, and anxiety. The stimulation by pathogens triggers the secretion of pro-inflammatory cytokines, including interleukins-1, 6, and 18, by various cellular players such as intestinal epithelial cells, dendritic cells, and macrophages. Notably, probiotics have demonstrated the potential to counteract these effects by orchestrating the activity of gut autonomous lymphoid tissue, maintaining an intricate equilibrium between pro- and anti-inflammatory cytokines (Cristofori et al. 2021, Hemarajata and Versalovic 2013, Mazziotta et al. 2023).

In a sumptuous amalgamation of findings, the study conducted by Erny et al. has unveiled a revolutionary insight into the harmonious interplay between the gut microbiota and microglial activation, effectively unravelling a cascading sequence of effects that resonates from the gut to the brain. This revelation holds immense potential in deepening our comprehension of the intricate nexus between neuro-immune interactions, while simultaneously ushering in novel avenues for interventions aimed at optimizing brain health through strategic modulation of the gut microbiota.

4.6 THE TRIAGE: MICROBIOTA AND NEURO- IMMUNE SYSTEM

The dynamic interplay between neurochemicals generated by both the host and gut microorganisms, including biogenic amines, amino acids, and SCFAs, serves as a pivotal mediator bridging the realms of gut microorganisms, the immune system, and the nervous system. This interaction assumes a bidirectional nature, wherein the low-molecular-weight compounds produced by microbes bear similarity or even identity to the intermediates synthesized by immune or nervous cells. Consequently, these compounds can effectively bind to corresponding receptors within the host system. Simultaneously, due to the intricate coevolutionary relationship between gut microbes and their host, these microorganisms have evolved to respond to neuromediators and neurohormones produced by the host, especially biogenic amines.

Interestingly, this coevolution has led to a remarkable scenario where the host relies on certain neuromediators to be synthesized by its microbial counterparts for the attainment of normal functioning. An illustrative example of this phenomenon is the synthesis of γ-aminobutyric acid (GABA) by microbes, which the host necessitates for its optimal operation (Otaru et al. 2021).

As previously mentioned, the influence of microbial metabolites transcends the boundaries of immune cells. These metabolites intricately interact with various physiological systems, including metabolic, epigenetic, and neuroendocrine systems. These multifaceted interactions hold the potential to induce notable modifications in the function of the nervous system (Figure 4.1).

Described below are the neurohormones/neuromediators making the triage (microbiota-immune-nervous) possible (reviewed in (Oleskin, Shenderov, and Rogovsky 2017)):

- Catecholamines- Catecholamines represent a class of neurotransmitters that are synthesized and secreted by various systems within the body. These compounds have been demonstrated to exert influence over the growth of several pathogenic microorganisms, such as *Yersinia enterocolitica, Escherichia coli, Shigella spp., Salmonella spp.*, and *Pseudomonas aeruginosa*. Notably, the levels of catecholamines experience an elevation during periods of stress, actively participating in the activation of the sympathetic nervous system. Within this context, both dopamine and norepinephrine, which fall under the category of catecholamines, assume the roles of hypothalamic neurohormone and adrenal hormone, respectively. These roles are intricately linked to pivotal cognitive processes, information retention, and emotional regulation.

  It's intriguing to note that immune cells are equipped with specialized receptors, primarily β2-adrenoreceptors (β AR), enabling them to detect the presence of catecholamines. Remarkably, these immune cells also possess the capacity to synthesize and release catecholamines themselves. The

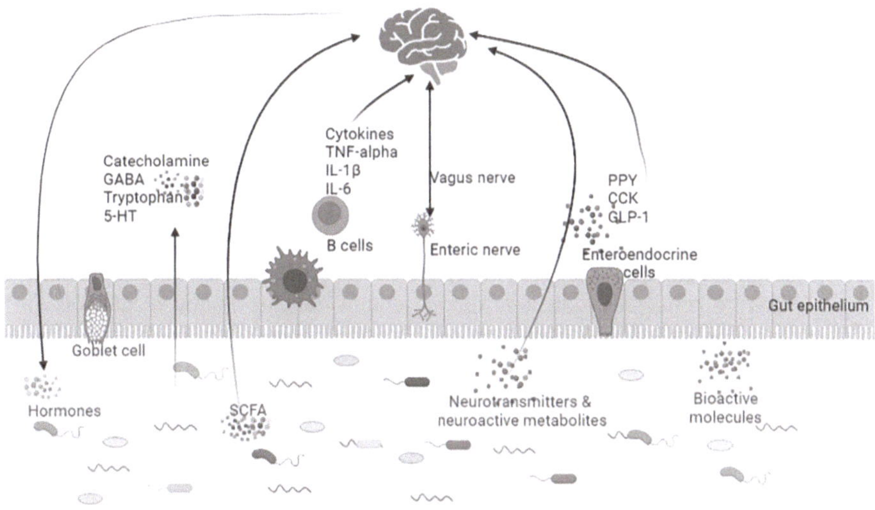

FIGURE 4.1 A comprehensive network of microbiota-gut-brain axis comprising intricate pathways transmitting signals from gut bacteria to the brain: These pathways involve various mechanisms, such as vagus nerve activation, immune mediator and microbial metabolite generation (including SCFAs), and enteroendocrine cell signaling. This axis exerts control over critical central physiological processes, encompassing neurotransmission, neurogenesis, neuroinflammation, and neuroendocrine signaling. Key signaling molecules, including serotonin (5-HT), cholecystokinin (CCK), γ-aminobutyric acid (GABA), glucagon-like peptide (GLP), interleukin (IL), peptide YY (PYY), tumor necrosis factor (TNF), and SCFAs, are pivotal in this intricate communication network. These pathways interact with those involving hepatic and gallbladder metabolism, immune-modulatory responses, neuronal connections, enteroendocrine signaling, and microbial metabolite transmission, forming a complex web of bidirectional interactions.

outcome of the interaction between catecholamines and specific receptors is highly context-dependent. For instance, this interaction can result in either the inhibition of the immune system and attenuation of inflammation, or conversely, the stimulation of immune responses. This distinction is particularly evident in the case of β AR and α AR receptors, where the former is associated with immunosuppression while the latter is linked to immune response enhancement.

Broadly, both dopamine and norepinephrine exhibit properties that extend beyond their neurotransmitter roles. They engage in a duality of functions as both modulators of neural activity and orchestrators of immune responses. It's notable that these catecholamines, especially dopamine and norepinephrine, display tendencies toward anti-inflammatory and immunosuppressive activities. This intricate interplay between catecholamines, immune cells, and neural processes showcases the complexity of the body's regulatory mechanisms, where the neurotransmitter landscape extends its influence far beyond the realms of neural communication to intricately shape immune responses and inflammatory states.

- Acetylcholine- Acetylcholine stands as a pivotal neuromediator, holding a significant role within both the sympathetic and parasympathetic nervous systems. Produced by a multitude of microorganisms, its receptors extend even to unicellular eukaryotes, where it has been observed to wield influence over essential processes like conjugation.

  The indispensable nature of acetylcholine extends beyond its role as a neuromediator, encompassing a spectrum of cognitive functions and neural activities. Central to attention span, memory, learning, motivation, plasticity, and overall brain activity, acetylcholine forms a cornerstone for these cognitive processes. The disruption of this delicate balance can consequently lead to the manifestation of characteristic memory loss observed in conditions like Alzheimer's disease.

  Notably, acetylcholine's significance isn't confined to the realms of cognition alone. It finds its relevance in the orchestration of skeletal muscle contractions, where it finds its abode in neuromuscular synapses, ensuring the smooth coordination of muscular movements.

  Acetylcholine, beyond its role in neurotransmission, surprisingly demonstrates anti-inflammatory attributes. This is exemplified by its interaction with nAchRα7 receptors, prevalent not only in various tissues but also in macrophages and other immune cells. Through this interaction, acetylcholine effectively inhibits the NF-κB pathway, a key regulator of pro-inflammatory cytokine production. This engagement underscores its potential as an anti-inflammatory agent, mitigating the excessive release of pro-inflammatory molecules and thereby contributing to immune regulation.

  Remarkably, the capabilities of acetylcholine extend to immune cells themselves. Both lymphocytes and macrophages are endowed with the ability to synthesize acetylcholine. This unexpected dimension unveils the intricate web of communication that exists between the nervous and immune systems, where neurotransmitters are not only messengers within the neural domain but also intricately woven into the immune system's regulatory framework.
- Serotonin- Serotonin, a multifaceted neurotransmitter, exhibits a range of roles that delve beyond its initial neurological functions. The majority of serotonin within the body originates from the metabolic breakdown of tryptophan, an amino acid commonly found in dietary sources. This intricate synthesis takes place within specialized endocrine cells nestled in the intestinal mucosa, establishing a unique link between neural signaling and gut physiology.

  Remarkably, serotonin's reach extends to microbial interactions. It serves as a growth enhancer for beneficial bacterial strains such as E. coli and Enterococcus faecalis, fostering a symbiotic relationship between neurotransmitters and the gut's microbial inhabitants. Conversely, selective serotonin uptake inhibitors, often employed to manage mental health conditions, exhibit the intriguing capacity to inhibit the growth of specific pathogens like *Candida parapsilosis* and mycelium fungi belonging to the *Aspergillus* genus.

Serotonin's influence doesn't halt there. It has been observed to mitigate the virulence of the pathogen *C. Albicans*, shedding light on its role in shaping microbial behavior and pathogenicity.

Intriguingly, the microbiota itself contributes significantly to serotonin production. This neurotransmitter, renowned for its involvement in emotional regulation, memory, learning, appetite, as well as social and sexual behaviors, emerges as a key player in the intestinal nervous system. Its scope goes beyond neural functions, extending to crucial roles in maintaining the integrity of the GI barrier and fostering food tolerance through the modulation of local inflammatory responses.

Serotonin's impact stretches to the immune domain as well. Immune cells are equipped with a diverse array of serotonin receptors and have the capability to produce their own serotonin. Notably, immune cells lacking the pivotal enzyme tryptophan hydroxylase, essential for serotonin synthesis, such as dendritic cells, can still store and release serotonin. This intricate interplay between immune and neural systems reveals a new layer of cross-talk between these systems.

Functionally, serotonin emerges as a pivotal orchestrator of immune responses. It significantly influences the proliferation, differentiation, and activity of immune cells, while navigating a delicate balance between stimulation and inhibition. This balance is governed by the intricate interplay of receptors, activation levels, and the unique microenvironment of each scenario.

- Histamine- Histamine, a multifunctional molecule, operates as both a neurotransmitter and a key player in local inflammatory responses, sharing this role with serotonin. While its presence spans various contexts, including both physiological and pathological conditions, histamine's production by certain microbial strains adds a layer of complexity to its impact.

  Remarkably, histamine is primarily synthesized by gram-negative pathogens like *Branhamella catarrhalis, Haemophilus parainfluenzae*, and *Ps. aeruginosa.* This microbial histamine production, intriguingly absent in gram-positive species, introduces an additional facet to the interplay between host and microorganisms. This becomes particularly significant in chronic respiratory conditions like chronic bronchitis or pneumonia, notably among individuals with heritable cystic fibrosis. In such cases, microbial histamine production could potentially exacerbate disease relapses, posing challenges in patient management.

  Histamine's role extends beyond its microbial interactions. It intricately influences various physiological processes, including pain sensitivity, appetite regulation, brain activity, and circadian rhythm. Within the CNS, histamine predominantly interacts with two types of G protein-coupled receptors, namely H1 and H3, contributing to a wide array of neural responses.

  In the realm of inflammation, histamine assumes a dualistic role. When bound to the H1 receptor, it exhibits pro-inflammatory behavior, potentially contributing to the amplification of inflammatory responses. However, a contrasting immunosuppressive effect comes into play when histamine binds to the H2 receptor. This interaction leads to the increased secretion of anti-inflammatory interleukin IL-10, adding a layer of immune regulation to histamine's repertoire.

Notably, the synthesis and storage of histamine are not limited to microbes. Immune cells, such as adipocytes and macrophages, are capable of producing and retaining histamine. This dynamic suggests that histamine acts as an immune modulator, potentially playing a role in shaping immune responses based on its presence and interactions with specific receptors.

- Agmatine- Agmatine, a multifaceted molecule with diverse interactions, holds sway over a range of physiological processes. Its intricate influence spans ion channels, NADPH oxidase, and the synthesis of nitric oxide (NO). Beyond these, its interactions with specific microorganisms unveil targeted effects, as evidenced by its role in suppressing intestinal colonization by the parasite *Cryptosporidium parvum*, thus thwarting potential infections.

  Operating as a neuromodulator and cotransmitter, agmatine exerts its impact on neurotransmitter systems. This is demonstrated through its binding to α2-adrenergic and imidazoline receptors, as well as its blocking of N-Methyl-D-Aspartate (NMDA) receptors. Notably, while agmatine's effects on other neurotransmitters are well-documented, the existence of dedicated agmatine receptors remains yet unverified.

  A pivotal role emerges in agmatine's influence on the conversion of arginine, a precursor amino acid, into two distinct compounds: NO and ornithine. NO, with its antimicrobial properties, serves as a defense against infectious agents, while ornithine fuels cell proliferation in inflammatory regions. Agmatine contributes to the anti-inflammatory milieu by inhibiting nitric oxide synthase (NOS), mitigating the inflammatory cascade.

  Remarkably, agmatine's influence extends to neuroprotection, particularly in circumstances of disrupted cerebral blood circulation. It showcases its prowess by reducing necrosis areas and conferring a vasoprotective effect on the brain, thereby mitigating damage. The molecular underpinning of its neuroprotective functions involves the activation of the nuclear transcription factor NRF2, which orchestrates cellular responses to oxidative stress. This activation, in turn, stimulates the production of antioxidant enzymes, reinforcing cellular defense mechanisms.

- Neuroactive amino acids (Lu et al. 2014)- Neuroactive amino acids, an intricate group encompassing both excitatory and inhibitory neurotransmitters, play a pivotal role in the intricate web of mood regulation. Within this spectrum, excitatory neurotransmitters like glutamic acid and aspartic acid hold the reins of neuronal activation, while their inhibitory counterparts, glycine and GABA, temper neural activity, fostering a delicate equilibrium.

  However, this balance can be perturbed in the context of mood disorders, particularly major depressive disorder (MDD). These amino acids are entangled in the complex pathogenesis of such disorders. Emerging evidence underscores their involvement in the delicate interplay of neuronal signaling that underpins mood regulation. Dysregulations in the levels and function of these neurotransmitters have been implicated in the manifestation of MDD.

Another intriguing player in this intricate symphony is the gaseous neurotransmitter NO. Synthesized from the amino acid L-arginine by the enzyme NOS, NO wields considerable influence over neural communication. In this process, citrulline is generated as a by-product. The multifaceted role of NO in depression is illuminated through its modulation of neurovascular coupling, cerebral blood flow, and neural plasticity. Its intricate involvement further underscores the complexity of the neural mechanisms implicated in mood regulation.

The intricate landscape of MDD isn't limited solely to neurochemical imbalances. It's a multifactorial interplay, where genetic, early developmental, and environmental factors converge to create a perfect storm. The intricate orchestration of this convergence can lead to a heightened responsiveness of an individual's stress response system to life's challenges. This system, normally designed to cope with stress, can go awry, pushing it into an overdrive mode. The result is the intricate tapestry of MDD, where neurotransmitters and their delicate equilibrium intersect with genetic predispositions and environmental triggers.

- SCFAs (Silva, Bernardi, and Frozza 2020) - SCFAs, the metabolic products resulting from the microbial fermentation of dietary fibers and resistant starch in the colon, are emerging as central players in the intricate symphony of neuro-immunoendocrine regulation. These molecules, speculated to possess a profound impact on brain function and behavior, are at the forefront of scientific inquiry. While their exact mechanisms remain enigmatic, recent studies underline their potential significance, particularly in the context of disorders spanning both the neurological and GI realms.

The foundation of this exploration lies in the dynamic interplay between the gut microbiota and its host. A growing body of evidence suggests that imbalances in intestinal bacterial communities contribute to the pathogenesis of diverse disorders. This interaction is a two-way street, with communication mechanisms abounding. Essential to this dialogue are the biochemical messengers crafted by the microbiota, unveiling their pivotal role in the crosstalk between these microbial communities and the host's physiological systems.

Foremost among these messengers are the SCFAs, which don multiple hats in shaping host well-being. Their influence spans a wide spectrum, encompassing GI physiology, immune response modulation, and even the nuanced intricacies of CNS development and homeostasis. This multifaceted impact stems from the ability of SCFAs to wield direct and indirect regulatory effects on various CNS processes. This orchestration, in turn, resonates outward, shaping behaviors and cognitive functions.

Unraveling the intricacies of how SCFAs participate in the complex interactions between the gut and the brain holds considerable promise for therapeutic advancements. Their potential to modulate CNS processes, potentially leading to the amelioration of disorders affecting the brain and behavior, opens new avenues for treatment strategies. This deeper understanding could serve as a cornerstone for novel therapeutic targets, fostering innovative interventions for CNS disorders.

Moreover, the influence of SCFAs extends beyond therapeutic realms, seeping into the realm of dietary interventions. Given their potential to impact the development and maintenance of healthy brain function, these metabolites hold the allure of being used as components of dietary approaches aimed at enhancing psychological well-being. As research delves deeper into the mechanisms and effects of SCFAs on CNS function, their potential utility in shaping psychological states becomes increasingly apparent.

4.7 CONCLUSION AND FUTURE DIRECTIONS

A harmonious relationship between the microbiota and Neuron–MM interactions is pivotal for maintaining optimal neuro-immune crosstalk. Conversely, when this balance is disrupted due to microbial dysbiosis, the intricate web of communication between neurons and the immune system can be negatively impacted. An illustrative example of this is the context of enteric infection-induced loss of enteric neurons, which is intricately linked to microbial dysbiosis resulting from the infection. Remarkably, the administration of antibiotics following pathogen clearance, followed by the recolonization with a healthy microbiota, triggers a significant recovery of enteric neurons, often restoring them close to pre-infection levels.

While the significance of such recovery is evident, a deeper understanding of how the microbiota orchestrates the neuron–MM signaling to drive or inhibit enteric neuron recovery is yet to be fully elucidated. The specific commensal species or microbial products involved in this process remain to be identified. Intriguingly, it has been observed that certain neuronal populations regulate the composition of the intestinal microbiota. For instance, sensory neuron-depleted mice demonstrate altered microbial communities, particularly evident in the reduced presence of segmented filamentous bacteria. This commensal's regulation by neurons mediates small intestine colonization resistance against infections like *Salmonella*.

The reversible nature of microbial modulation further underscores its dynamic influence on neuro-immune interactions. As showcased in various instances, sequential modifications of the microbiota can reshape signaling mechanisms. However, this field is ripe for more extensive exploration, as the examples presented might only scratch the surface of these intricate interactions. Unraveling the full spectrum of neuro-immune interactions mediated by the microbiota remains a compelling challenge.

To delve deeper into this enigma, innovative tools and approaches are being harnessed. Advanced techniques like optogenetics, chemogenetics, and imaging, coupled with analysis of immune cell populations, hold the potential to unveil novel insights into the realm of neuro-immune crosstalk. Furthermore, the reciprocal relationship between gut-resident immune cells and neurons is yet another fascinating area of inquiry. Employing techniques like calcium imaging and neural tracing, researchers are striving to comprehend how immune cells communicate with neurons and how this communication influences immune responses.

This evolving field also promises insights into the realm of disease and potential therapeutic avenues. Dysregulated neuro-immune circuitry has been identified in various inflammatory disease contexts. The prospect of bioelectric medicine, such as targeted electrical stimulation of nerves like the vagus nerve, is gaining traction as a potential treatment strategy for inflammatory diseases. This trend is likely to continue, potentially leading to transformative therapeutic approaches for conditions like IBD, enteric infections, and food allergies.

Ultimately, the intricate relationship between the nervous and immune systems emerges as a symphony of coordinated responses to environmental cues. These systems, once considered separately, are now recognized as integral components of an integrated system. The ongoing mechanistic studies exploring neuro-immune signaling promise to unveil a deeper understanding of this integrated system and its profound impact on the generation and maintenance of immunity. As research progresses, the next few years are poised to yield remarkable strides in uncovering the complexity of this interconnected web and its implications for health and disease.

APPENDIX-1

Terminology

GI	gastrointestinal
ENS	enteric nervous system
PRRs	pattern recognition receptors
PAMPs	pathogen-associated, olecular patterns
LPS	lipopolysaccharide
ECs	enteroendocrine or enterochromaffin cells
TLRs	toll-like receptors
MM	muscularis macrophages
CNS	centralnervous system
MHC	major histocompatibility complex
BBB	blood brain barrier
Tregs regulatory	T cells
SCFAs	short-chain fatty acids
IBD	inflammatory bowel disease
5-HT	serotonin
CCK	cholecystokinin
GABA	γ-aminobutyric acid
GLP	glucagon-like peptide
IL	interleukin
PYY	peptide YY
TNF	tumor necrosis factor
NO	nitric oxide
NMDA	N-Methyl-D-Aspartate
NOS	nitric oxide synthase
MDD	major depressive disorder

REFERENCES

Bohorquez, D. V., R. A. Shahid, A. Erdmann, A. M. Kreger, Y. Wang, N. Calakos, F. Wang, and R. A. Liddle. 2015. "Neuroepithelial circuit formed by innervation of sensory enteroendocrine cells." *J Clin Invest* 125 (2): 782–6. doi: 10.1172/JCI78361

Braniste, V., M. Al-Asmakh, C. Kowal, F. Anuar, A. Abbaspour, M. Toth, A. Korecka, N. Dakocevic, L. G. Ng, P. Kundu, B. Gulyas, C. Halldin, K. Hultenby, H. Nilsson, H. Hebert, B. T. Volpe, B. Diamond, and S. Pettersson. 2014. "The gut microbiota influences blood-brain barrier permeability in mice." *Sci Transl Med* 6 (263): 263ra158. doi: 10.1126/scitranslmed.3009759

Brown, D. G., R. Soto, S. Yandamuri, C. Stone, L. Dickey, J. C. Gomes-Neto, E. D. Pastuzyn, R. Bell, C. Petersen, K. Buhrke, R. S. Fujinami, R. M. O'Connell, W. Z. Stephens, J. D. Shepherd, T. E. Lane, and J. L. Round. 2019. "The microbiota protects from viral-induced neurologic damage through microglia-intrinsic TLR signaling." *Elife* 8. doi: 10.7554/eLife.47117

Cristofori, F., V. N. Dargenio, C. Dargenio, V. L. Miniello, M. Barone, and R. Francavilla. 2021. "Anti-inflammatory and immunomodulatory effects of probiotics in gut inflammation: A door to the body." *Front Immunol* 12:578386. doi: 10.3389/fimmu.2021.578386

Dalile, B., L. Van Oudenhove, B. Vervliet, and K. Verbeke. 2019. "The role of short-chain fatty acids in microbiota-gut-brain communication." *Nat Rev Gastroenterol Hepatol* 16 (8):461–478. doi: 10.1038/s41575-019-0157-3

Dey, N., V. E. Wagner, L. V. Blanton, J. Cheng, L. Fontana, R. Haque, T. Ahmed, and J. I. Gordon. 2015. "Regulators of gut motility revealed by a gnotobiotic model of diet-microbiome interactions related to travel." *Cell* 163 (1): 95–107. doi: 10.1016/j.cell.2015.08.059

Diaz Heijtz, R., S. Wang, F. Anuar, Y. Qian, B. Bjorkholm, A. Samuelsson, M. L. Hibberd, H. Forssberg, and S. Pettersson. 2011. "Normal gut microbiota modulates brain development and behavior." *Proc Natl Acad Sci U S A* 108 (7): 3047–52. doi: 10.1073/pnas.1010529108

Erny, D., A. L. Hrabe de Angelis, D. Jaitin, P. Wieghofer, O. Staszewski, E. David, H. Keren-Shaul, T. Mahlakoiv, K. Jakobshagen, T. Buch, V. Schwierzeck, O. Utermohlen, E. Chun, W. S. Garrett, K. D. McCoy, A. Diefenbach, P. Staeheli, B. Stecher, I. Amit, and M. Prinz. 2015. "Host microbiota constantly control maturation and function of microglia in the CNS." *Nat Neurosci* 18 (7): 965–77. doi: 10.1038/nn.4030

Gribble, F. M., and F. Reimann. 2019. "Function and mechanisms of enteroendocrine cells and gut hormones in metabolism." *Nat Rev Endocrinol* 15 (4): 226–237. doi: 10.1038/s41574-019-0168-8

Hemarajata, P., and J. Versalovic. 2013. "Effects of probiotics on gut microbiota: mechanisms of intestinal immunomodulation and neuromodulation." *Therap Adv Gastroenterol* 6 (1): 39–51. doi: 10.1177/1756283X12459294

Lu, Y. R., X. Y. Fu, L. G. Shi, Y. Jiang, J. L. Wu, X. J. Weng, Z. P. Wang, X. Y. Wu, Z. Lin, W. B. Liu, H. C. Li, J. H. Luo, and A. M. Bao. 2014. "Decreased plasma neuroactive amino acids and increased nitric oxide levels in melancholic major depressive disorder." *BMC Psychiatry* 14: 123. doi: 10.1186/1471-244X-14-123

Marklund, U. 2022. "Diversity, development and immunoregulation of enteric neurons." *Nat Rev Gastroenterol Hepatol* 19 (2): 85–86. doi: 10.1038/s41575-021-00553-y

Maslowski, K. M., A. T. Vieira, A. Ng, J. Kranich, F. Sierro, D. Yu, H. C. Schilter, M. S. Rolph, F. Mackay, D. Artis, R. J. Xavier, M. M. Teixeira, and C. R. Mackay. 2009. "Regulation of inflammatory responses by gut microbiota and chemoattractant receptor GPR43." *Nature* 461 (7268): 1282–6. doi: 10.1038/nature08530

Mazziotta, C., M. Tognon, F. Martini, E. Torreggiani, and J. C. Rotondo. 2023. "Probiotics mechanism of action on immune cells and beneficial effects on human health." *Cells* 12 (1). doi: 10.3390/cells12010184

Muller, P. A., B. Koscso, G. M. Rajani, K. Stevanovic, M. L. Berres, D. Hashimoto, A. Mortha, M. Leboeuf, X. M. Li, D. Mucida, E. R. Stanley, S. Dahan, K. G. Margolis, M. D. Gershon, M. Merad, and M. Bogunovic. 2014. "Crosstalk between muscularis macrophages and enteric neurons regulates gastrointestinal motility." *Cell* 158 (5): 1210. doi: 10.1016/j.cell.2014.08.002

Ogbonnaya, E. S., G. Clarke, F. Shanahan, T. G. Dinan, J. F. Cryan, and O. F. O'Leary. 2015. "Adult Hippocampal Neurogenesis Is Regulated by the Microbiome." *Biol Psychiatry* 78 (4): e7–9. doi: 10.1016/j.biopsych.2014.12.023

Oleskin, A. V., B. A. Shenderov, and V. S. Rogovsky. 2017. "Role of neurochemicals in the interaction between the microbiota and the immune and the nervous system of the host organism." *Probiotics Antimicrob Proteins* 9 (3): 215–234. doi: 10.1007/s12602-017-9262-1

Otaru, N., K. Ye, D. Mujezinovic, L. Berchtold, F. Constancias, F. A. Cornejo, A. Krzystek, T. de Wouters, C. Braegger, C. Lacroix, and B. Pugin. 2021. "GABA production by human intestinal bacteroides spp.: Prevalence, regulation, and role in acid stress tolerance." *Front Microbiol* 12: 656895. doi: 10.3389/fmicb.2021.656895

Parada Venegas, D., M. K. De la Fuente, G. Landskron, M. J. Gonzalez, R. Quera, G. Dijkstra, H. J. M. Harmsen, K. N. Faber, and M. A. Hermoso. 2019. "Short Chain Fatty Acids (SCFAs)-mediated gut epithelial and immune regulation and its relevance for inflammatory bowel diseases." *Front Immunol* 10: 277. doi: 10.3389/fimmu.2019.00277

Sanidad, K. Z., and M. Y. Zeng. 2020. "Neonatal gut microbiome and immunity." *Curr Opin Microbiol* 56: 30–37. doi: 10.1016/j.mib.2020.05.011

Shigemoto-Mogami, Y., and K. Sato. 2021. "[Central nervous system developmental regulation of microglia via cytokines and chemokines]." *Yakugaku Zasshi* 141 (3): 359–368. doi: 10.1248/yakushi.20-00198-4

Silva, Y. P., A. Bernardi, and R. L. Frozza. 2020. "The role of short-chain fatty acids from gut microbiota in gut-brain communication." *Front Endocrinol (Lausanne)* 11: 25. doi: 10.3389/fendo.2020.00025

Smith, P. M., M. R. Howitt, N. Panikov, M. Michaud, C. A. Gallini, Y. M. Bohlooly, J. N. Glickman, and W. S. Garrett. 2013. "The microbial metabolites, short-chain fatty acids, regulate colonic Treg cell homeostasis." *Science* 341 (6145): 569–73. doi: 10.1126/science.1241165

Stilling, R. M., T. G. Dinan, and J. F. Cryan. 2014. "Microbial genes, brain & behaviour - epigenetic regulation of the gut-brain axis." *Genes Brain Behav* 13 (1): 69–86. doi: 10.1111/gbb.12109

Trompette, A., E. S. Gollwitzer, K. Yadava, A. K. Sichelstiel, N. Sprenger, C. Ngom-Bru, C. Blanchard, T. Junt, L. P. Nicod, N. L. Harris, and B. J. Marsland. 2014. "Gut microbiota metabolism of dietary fiber influences allergic airway disease and hematopoiesis." *Nat Med* 20 (2): 159–66. doi: 10.1038/nm.3444

Veiga-Fernandes, H., and D. Mucida. 2016. "Neuro-immune interactions at barrier surfaces." *Cell* 165 (4): 801–11. doi: 10.1016/j.cell.2016.04.041

Wang, H., J. P. P. Foong, N. L. Harris, and J. C. Bornstein. 2022. "Correction to: Enteric neuroimmune interactions coordinate intestinal responses in health and disease." *Mucosal Immunol* 15 (1): 188. doi: 10.1038/s41385-021-00459-7

Yano, J. M., K. Yu, G. P. Donaldson, G. G. Shastri, P. Ann, L. Ma, C. R. Nagler, R. F. Ismagilov, S. K. Mazmanian, and E. Y. Hsiao. 2015. "Indigenous bacteria from the gut microbiota regulate host serotonin biosynthesis." *Cell* 161 (2): 264–76. doi: 10.1016/j.cell.2015.02.047

5 Neuroimmune Biology in Cancer

Shyamasree Ghosh
National Institute of Science Education and Research (NISER), Bhubaneswar, an OCC of Homi Bhabha National Institute (HBNI), Bhubaneswar, India

Sanjima Pal
Research Institute-McGill University Health Centre (RI-MUHC), Canada

Suchandra Chowdhury
Post Graduate Department of Zoology, Bethune College, 181, Bidhan Sarani, Kolkata, India

5.1 INTRODUCTION

The intricate interplay between the immune and nervous systems within the tumor microenvironment is essential for coordinating molecular and pathophysiological responses during the initiation, progression, or remission of tumors. This intricate communication network involves the release of cytokines, chemokines, and growth factors, creating a milieu that can either promote or inhibit tumor growth. In 1987, early investigations by Faith and colleagues marked a significant milestone in neuroimmunology, demonstrating the regulatory role of neurotransmitters, particularly enkephalins, in immune function among cancer patients (Faith et al., 1987). At that time, the immune system was largely perceived as a self-governing defence mechanism, operating independently of neural control through cellular interactions (Ader, 1995). However, a transformative paradigm shift occurred as subsequent studies illuminated the close collaboration between the immune system and the nervous system in disease pathogenesis. These findings unveiled a previously unrecognized connection, highlighting the integral role of the nervous system in regulating inflammation and orchestrating host responses to injury and tumors (Abramsky et al., 1987, Barni et al. 1988, Dantzer, 2018; Scheff et al., 2021).

Immune cells engage in dynamic interactive bidirectional communication with neurons and tumor cells through diverse signaling pathways. This process encompasses communication For more than a decade now, it has been clear that both the adaptive and innate immune systems function in harmony with the nervous system. Evidences indicate that the Central Nervous System (CNS) produces and utilizes

 DOI: 10.1201/9781003295471-5

immune factors, while the immune system produces and utilizes neuroendocrine mediators. The nervous system, through its communication with immune cells, can either promote or suppress tumor growth and metastasis. Understanding the complexities of neuroimmune crosstalk is essential for unraveling the mechanisms that contribute to cancer progression, paving the way for novel therapeutic strategies that target both the nervous and immune systems to effectively combat tumorigenesis. Immune cells and nerve endings in immune organs interact, utilizing local neuronal components to finely tune the immune response. The regulatory role of immune cells and their mediators in the nervous system leads to the removal and plasticity of synapses during development and in synaptic plasticity. Long-distance interactions between the CNS and immune cells empower the immune system to combat pathogenic infections, while simultaneously enabling the nervous system to regulate immune system functions (Dantzer et al., 2018).

Lymphoid structures, along with tumors closely associated with inflammatory infiltrate, are innervated by various types of nerves, including sympathetic, parasympathetic, and sensory nerves (Scheff et al., 2021). New discoveries regarding the role of the CNS and both the peripheral nervous system (PNS) and CNS in regulating tumorigenesis and metastasis are bringing attention to innovative strategies for the targeting and treatment of cancer. Cancer prognosis has recently been associated directly within intra-tumoral neural infiltration, arising in highly innervated organs like pancreas cancer, head and neck cancers, prostate and colorectum cancers (Bapat et al., 2011; Silverman et al., 2021; Ayala & Chen, 2018). The nervous system directed interventions not only has the potential to enhance therapeutic strategies for promoting anti-tumor immunity but also to alleviate cancer-associated neurological symptoms such as chronic pain, fatigue, and cognitive impairment. (Scheff et al., 2021). The neuroimmune crosstalk in tumor development will be the main focus of this chapter.

5.2 HISTORICAL RETROSPECTIVE

The intricate interplay between the nervous and immune systems, known as neuroimmune crosstalk, plays a crucial role in disease pathogenesis. Neuroimmune crosstalk influences key aspects of tumorigenesis, including the regulation of inflammation, immune responses, and the establishment of a tumor-supportive microenvironment. Additionally, neuroimmune crosstalk may contribute to the evasion of immune surveillance by tumors, allowing them to establish and maintain a foothold within the host. Understanding these mechanisms is essential for developing targeted therapeutic interventions aimed at disrupting the neuroimmune crosstalk that fuels tumor promotion, potentially offering new avenues for cancer treatment.

The initial information of nerves infiltrating tumors emerged in the beginning of the 20th century. The sensitivity and instinctive pain in certain tumors suggested the intricate roles played by new nerve fibres (Young 1897). Nerves have been reported in connection with the blood vessels, stroma, and the cellular parenchyma of the tumor (Oertel 1928).

In a classic experiment, mouse spinal nerves were placed alongside human prostate cancer cells *in vitro*. The findings revealed a mutual attraction between the nerves and cancer cells, as the nerves extended toward the expanding centers of the tumor cell colony. This phenomenon ultimately led to highlighting the dynamic interaction between nerves and cancer cells in a controlled laboratory setting. This also offered insights into the dynamic relationship between the nervous system and cancer progression. (Ayala et al., 2001). In 2006, Sood and his team implemented drug induced chronic stress in mice and demonstrated that this led to the activation of β-adrenergic receptors and subsequently resulted in increased growth of ovarian tumors. (Thaker et al., 2006). The ablation of adrenergic sympathetic and cholinergic parasympathetic nerves innervating prostate tumors has been reported to impair *in vivo* tumorigenesis (Magnon et al., 2023). Gustavo Ayala studied interactions between nerves and tumor cells and revealed that the interaction could lead to new therapies (Servick 2019). Over the past decade, research from various laboratories has unveiled the significant role of the nervous system in both initiating and advancing cancer. (Reavis et al., 2020).

5.3 NEUROIMMUNOLOGY OF CANCER-EMERGING CONCEPTS

The prevailing hypothesis driving tumorigenesis suggests that cancer cells manipulate immunological processes to create an immunosuppressive environment, allowing them to evade immune surveillance. Many tumors achieve this evasion by diminishing the population of effector cells and promoting the presence of myeloid-derived suppressor cells (MDSCs). Nerves contribute significantly to the regulation of tumor growth by fostering an pro-tumorigenic immune response. (Ribatti 2017; Scheff et al., 2021).

The tumor microenvironment (TME) encompasses not only malignant cells but also diverse subsets of effector and regulatory immune cell populations, along with the PNS consisting of sympathetic, parasympathetic, and sensory nerves. A significant interplay exists between the PNS and immune cells within the tumor, including macrophages, MDSCs, monocytes, neutrophils, and lymphocytes, such as natural killer (NK) cells and cytotoxic T cells. The modulation of neurotransmission through tailored interventions alters the activity of various immune targets. This cell-cell communication occurs either directly or is facilitated by immunoregulatory molecules such as chemokines, cytokines, growth factors, and checkpoint proteins. (e.g., tumor necrosis factor alpha (TNF-α), CTLA4, programmed cell death ligand-1 (PDL1) and programmed cell death-1 (PD1). Neurotransmitters and neuropeptides present within the TME play a regulatory role in cancer progression. They exert direct effects on malignant cells and tumor-infiltrating immune cell function. The blocking or activation of neurotransmission can elicit either pro-tumor or anti-tumor immune reactions in a nerve-type and tumor type-dependent manner (Scheff et al., 2021).

Sympathectomy, the surgical removal or functional disruption of sympathetic nerve fibers, has significant implications for tumor-related processes. Sympathectomy is linked to a systemic alteration in the neutrophil-to-lymphocyte ratio, accompanied by an elevated level of pro-inflammatory TNF-α. This suggests a potential augmentation in the extra-tumoral mobilization of innate immune cells, indicating the influence of sympathetic nervous system activity on systemic immune responses in

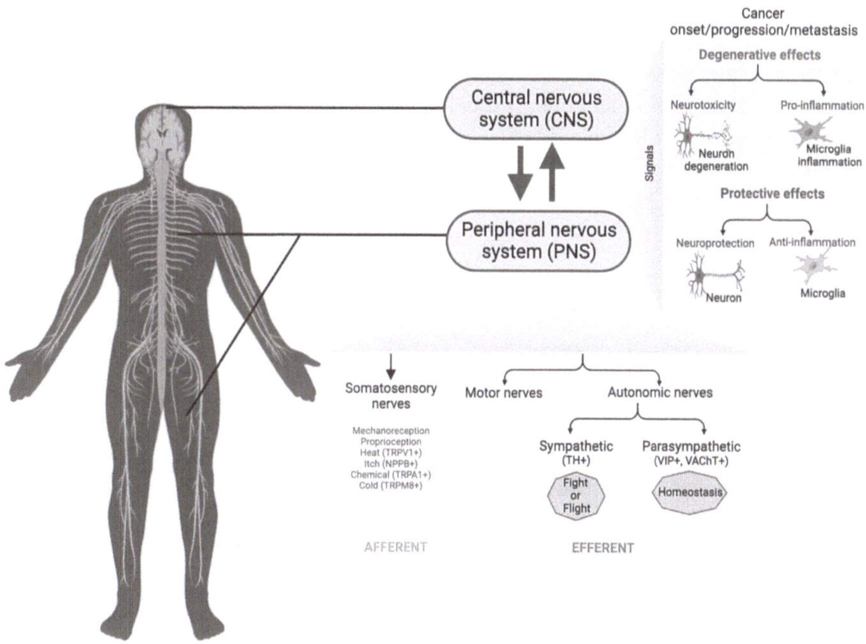

FIGURE 5.1 Overview of the human nervous system In a typical human, the central nervous system (CNS), comprising the brain and spinal cord, processes incoming signals from the peripheral nervous system (PNS) and generates responses through outgoing signals. The PNS includes somatosensory nerves that respond to stimuli from mechanoreceptors, proprioceptors, and other sensory receptors. The outgoing signals, or efferent signals, consist of both motor and autonomic nerves. Sympathetic nerves, responsible for "fight/flight" responses, and parasympathetic nerves, characterized by vasoactive intestinal peptide/vesicular acetylcholine transporter (VAChT+/VIP+), work in concert to maintain the body's internal balance and homeostasis. (Reavis et al., 2020). Signals received and processed by the CNS and PNS can result in the transmission of either neurodegenerative or neuroprotective signals during the pathogenesis of diseases.

the context of tumors. (Horvathova et al., 2015). Blockade of the beta-adrenergic signaling pathway is associated with a significant decrease in infiltration of MDSCs and an associated increase in cytotoxic lymphocytes population (Jean Wrobel et al., 2016) (Figure 5.1).

5.4 STRESS-SYMPATHETIC NERVOUS SYSTEM (SNS) AND TUMORIGENESIS AXIS

The SNS and hypothalamic–pituitary–adrenal axis jointly regulate the body's stress response. SNS's primary neurotransmitters, epinephrine and norepinephrine, bind to neuronal adrenergic receptors, initiating the "fight or flight" response. Immune cells expressing these receptors enable effective crosstalk, facilitating a regulatory system between the nervous and immune systems. (Sharma & Farrar, 2020; Scheff et al., 2021).

Experimental evidence has shown that stress-induced activation of the sympathetic nervous system (SNS) can alter the TME. Stress expedites immunotolerance by activating the hypothalamic-pituitary-adrenal axis, which, in turn, diminishes the cytotoxicity of natural killer cells (NK cells) and weakens cellular immune responses to tumor antigens. (Cole et al., 2015). Controlling the stress response either by reducing sympathetic nerve activity or promoting parasympathetic activity leads to the activation of effector T cells and the suppression of MDSCs within the tumor. (Scheff et al., 2021).

5.5 PERINEURAL INVASION

Perineural invasion (PNI), also known as perineural spread or neurotropic carcinomatous spread, is characterized by the infiltration of cancerous tissue by neoplastic cells into nerves. PNI represents a symbiotic relationship between cancer cells and nerves, marking a highly efficient interaction between these two cell types. This communication fosters a growth advantage for both cell types. In addition to the traditionally acknowledged vascular and lymphatic routes of metastasis, nerves constitute a third route for the spread of cancer. (Ayala et al., 2008; Ayala & Chen, 2018; Liebig et al., 2009)

In the early nineteenth century, PNI was first described in the European literature by scientists in the context of head and neck cancers (Cruveilheir 1835; Neumann 1862). The regulation and functioning of PNI involve various factors, including neurotransmitters, neuromodulators, immunoregulatory molecules (such as cytokines and chemokines), metabolic processes (e.g., serine metabolism), and other cellular mechanisms that encompass stromal cells, Schwann cells, T cells, and macrophages. This intricate combination of PNI mechanisms supports the development, growth, invasion, and metastasis of tumors. PNI also plays a role in inducing cancer-related pain, with this pain being correlated with a poorer disease prognosis. Instances of PNI have been documented in various cancers, and its presence is often associated with aggressive tumors and an unfavorable prognosis. (Li et al., 2021; Servick 2019)

5.5.1 Molecular Mediators of PNI

Neurotrophins, released by both neuronal and cancer cells, constitute a group of proteins that facilitate the survival, growth, and function of neurons. The expression of the nerve growth factor family (NGF) in pancreatic cancer cell lines suggests a potential interaction between these cancer cells and neural cells through the influence of neurotrophins. (Miknyoczki et al., 1999).

Glial cell-derived neurotrophic factor (GDNF) family ligands are a series of proteins which can bind to specific receptors to perform distinct functions like the maintenance of neuron growth and is highly expressed in the CNS and PNS. Results indicate that pancreatic cancer patients with PNI had a higher frequency of GDNF expression, which is involved in the neural invasion process (Airaksinen & Saarma, 2002; Zeng et al., 2008). The midkine (MDK) family promotes the outgrowth of neural cells. A high level of MDK in pancreatic cancer is associated with PNI and poor prognosis (Yao et al., 2014). Certain members of the axon guidance gene family, like semaphorin 3D (SEMA3D), contribute to the formation of neuronal networks. In instances where the expression of SEMA3D protein or its receptor is diminished, there is a corresponding reduction in nerve density within tumor tissues in vivo. (Jurcak et al., 2019). Catecholamines are the neurotransmitters containing epinephrine, norepinephrine, and dopamine and are important for maintaining unique physiological functions and stress responses. Norepinephrine induced tumor invasion to the nerves by activating multiple signaling pathways like the adrenoceptor beta 2 (ADRB2) (Guo et al., 2013). Excessive production of chemokines that primarily function as chemoattractants for both immune and neural cells is frequently linked to PNI (Xu et al., 2015; Li et al., 2021).

5.5.2 Cellular Mediators of PNI

The Schwann cells play a major role in nerve repair and regeneration. The interaction between the Schwann cells and cancer cell through neural cell adhesion molecule 1 (NCAM1) promote PNI both *in vitro* and *in vivo* (Deborde et al., 2016). Schwann cells that express and release L1 cell adhesion molecules (L1CAM) exhibit strong chemotaxis toward cancer cells (Na'ara et al., 2019).

Cancer-associated fibroblasts (CAFs) are stromal cells that undergo significant differentiation, promoting tumor growth, angiogenesis, and matrix remodelling. Nectin cell adhesion molecule 1 (NECTIN1) expression in CAFs is positively correlated with PNI (Yamada et al., 2018). The elevated number and density of immunosuppressive tumor-associated macrophages (TAMs) in a PNI-positive group indicates potential involvement in the development of PNI (Zeng et al., 2014; Li et al., 2021).

5.6 CANCER-INDUCED AXONOGENESIS

Cancer reinforces the connectivity between nerves and normal epithelium, a regulatory mechanism fundamental to the basic biology of epithelial cells in both benign and malignant conditions. In cancer, axonogenesis or axon morphogenesis refers to the growth of axonal projections from a neuronal cell, commonly known as neurite outgrowth or neurite sprouting. This process is attributed to the heightened volume of epithelial tissue that accompanies carcinogenesis. Therefore, induced axonogenesis is an acute phase phenomenon occurring within the PNS. Hypotheses suggest that axonogenesis/neurogenesis is followed by PNI and subsequent carcinogenic stages. The newly formed axons provide support for the newly generated malignant epithelial cells, contributing to cancer growth. Elevated axon density and neurite count have been reported in cancer, showing an association with increased aggression. (Ayala et al., 2008; 2023). This process is also found to be stimulated by the release of neurotrophic growth factors by cancer cells and cancer derived extracellular vehicles (EVs) containing altered micro-RNA (miRNA) levels (Silverman et al., 2021; Schmitd et al., 2021).Axonogenesis is clinically significant as a potential biomarker. Axonogenesis assists PNI, which becomes the second step of neuro-epithelial interactions. (Ayala et al., 2008). Quantification of nerve density using anti-protein gene product 9.5 (PGP 9.5) antibodies showed that higher nerve density correlated positively with higher recurrence in prostate cancer patients (Olar et al., 2014).

Numerous cancer cells express neurotrophic markers, such as brain-derived neurotrophic factor (BDNF), nerve growth factor (NGF), and glial cell–derived neurotrophic factor (GCDNF). Additionally, they release axon-guidance molecules, including ephrin B1, to facilitate axonogenesis. (Silverman et al., 2021). Therefore, a thorough understanding of the mechanisms underlying carcinoma-nerve interactions is crucial for the development of cancer-related therapies.

5.7 NEUROGENESIS

Neurogenesis succeeds axonogenesis and initially occurs in ganglia surrounding organs or the spinal column, later involving the recruitment of neuroblasts from the CNS. This process entails the formation of new neurons from neural precursor cells. Experimental findings indicate that the overexpression of semaphorin 4F (S4F) in cancer cells stimulates neurogenesis, while the inhibition of S4F by small interfering RNA (siRNA) hinders this process (Ayala et al., 2008; 2023).

A neuronal progenitors biomarker Doublecortin (DCX), demonstrated a significant association between high intratumoral expression and tumor aggressiveness in prostate cancer. Furthermore, it was observed that these cells migrate from the subventricular zone of the brain to the TME, suggesting a distant communication process that recruits neurons essential for tumorigenesis (Mauffrey et al., 2019; Schmitd et al., 2021).

5.8 NEUROTROPISM

The phenomenon of cancer cells migrating towards nerves is termed neurotropism. It has been noted that cancer cells exhibiting elevated expression of GALR2 (galanin receptor 2) demonstrated increased invasiveness toward the dorsal root ganglion (DRG), compared to control cells. Importantly, this effect could be inhibited by the administration of an anti-galanin antibody (Scanlon et al., 2015; Schmitd et al., 2021).

5.9 NEURAL REPROGRAMMING

Recent research suggests that newly formed adrenergic nerve fibers play a role in promoting tumor growth, but the origin and mechanism of their development remain unknown. Existing neural cells often undergo minor to major alterations or changes including gene expression and functionality, contributing to cancer development or progression (Silverman et al., 2021). This process may include factors such as changes in neural cell behaviour, migration, or interaction with cancer cells, influencing the TME and promoting tumorigenesis. Understanding neural reprogramming in cancer is crucial for developing targeted therapeutic strategies aimed at disrupting or reversing these neural changes to impede cancer growth and metastasis. Moran A et al. identified an adrenergic differentiation signature through a comparative transcriptomic study between cancer-associated trigeminal sensory neurons and endogenous neurons in mouse models of oral cancer. The research team further demonstrated that the loss of p53 is responsible for driving neuron reprogramming in head and neck cancer. Specifically, the loss of TP53 results in adrenergic

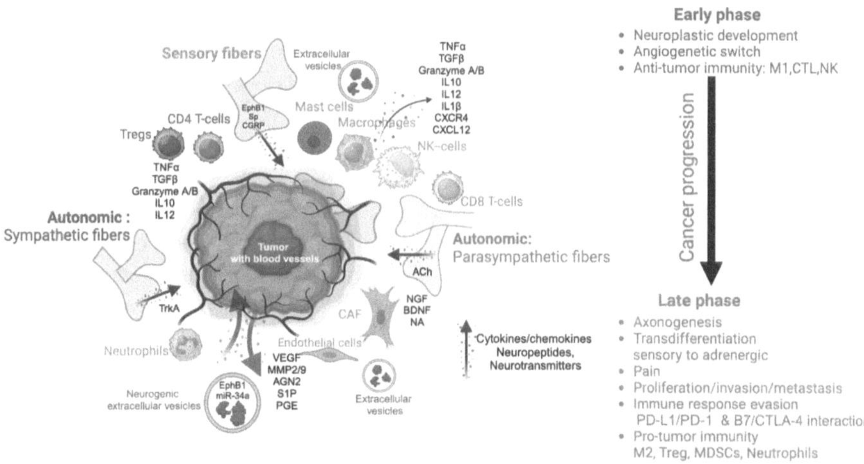

FIGURE 5.2 Components influencing the tumor-neuroimmune network in the TME. Neuroimmune crosstalk in tumor microenvironment is crucial for cancer immune responses. Immune cells, including macrophages, lymphocytes, and dendritic cells, communicate with neurons and tumor cells through various pathways, releasing cytokines, chemokines, and growth factors. The nervous system (both sympathetic and parasympathetic nervous systems) regulates immune cell function and infiltration, while sensory nerves contribute to inflammation and angiogenesis. Bidirectional communication between neurons and tumor cells can promote survival or alter neural function. Understanding the neuroimmune crosstalk in cancer holds significant clinical implications.

transdifferentiation of tumor-associated sensory nerves, a process facilitated by the downregulation of the microRNA miR-34a (Moran et al., 2020) (Figure 5.2)

5.10 ORIGIN OF NERVES ASSOCIATED WITH TUMOR

Cancer research has been directed toward understanding the intricate molecular mechanism involved in nervous and immune system interactions as well as psychophysiological reactions of the body to tumor invasion, facilitated by brain and body mechanisms, encompassing the endocrine, neuroimmune, and nervous system (Taams 2019).

A recent term, "neural addiction," has been introduced in the context of cancer, yet its full comprehension remains incomplete (Magnon et al., 2023). The involvement of peripheral and central nerve circuitries, along with brain areas, has been unveiled in supporting both tumorigenesis and metastasis, establishing a reciprocal influence between the brain and peripheral tumors.

Tumors have been found to establish local autonomic and sensory nerve networks, maintaining communication with the brain through circulating adipokines, inflammatory cytokines, neurotrophic factors, or afferent nerve inputs. This interaction promotes cancer initiation, growth, and dissemination. The CNS plays a significant role in tumor development and metastasis by either activating or dysregulating specific central neural areas or circuits, as well as influencing the neuroendocrine, neuroimmune, or neurovascular systems. Exploring and comprehending the interplay between brain neural circuitries and tumors, understanding brain-tumor communication, and delving into the interaction of intratumor nerves within the TME could unveil novel insights into the mechanisms underlying the initial phases of cancer development and progression. Targeting the dysregulated peripheral and central nervous systems (CNS) may represent a crucial step forward in cancer treatment (Magnon et al., 2023).

Cell-cell communication is integral to immune system function, facilitating precise and adjusted immune responses. Various components, such as cell surface and transmembrane receptors, ligands, junctions, membrane vesicles, ligands, voltage-gated ion channels, and transporters, actively participate in this communication within the immune system.

The interplay between the immune and nervous systems, central to cellular communication and signal transduction, is significantly impacted by neurotransmitters acting locally through autocrine or paracrine modes. This influence extends to shaping both innate and adaptive immune responses. Research highlights the crucial role of brain cytokines in mediating signals between immune cells and the brain, thereby influencing neural responses. Anomalies or disruptions in the communication between the nervous and immune systems are implicated in the onset of neurodegenerative and immunopathological diseases, encompassing autoimmune conditions and cancer. Neurotransmitters, including dopamine (3,4-dihydroxyphenethylamine), serotonin (5-hydroxytryptamine), substance P (trifluoroacetate salt powder), and L-glutamate, are recognized for exerting both immunostimulatory and immunosuppressive effects on lymphocytes and non-lymphoid immune cells. Understanding this neuroimmune crosstalk emphasizes the importance of therapeutic modulation to achieve a balanced interaction between the nervous and immune systems and holds significant implications for potential clinical interventions (Hodo et al., 2020).

Acute pain serves as an indicator of tissue damage with a protective function, while chronic pain may persist beyond the initial injury. Substantial alterations take place in peripheral and central pathways, resulting in a sensitized pain system characterized by heightened responses to noxious stimuli (hyperalgesia) and reactions to non-noxious stimuli (allodynia). Imaging studies have identified glial activation in the brain during chronic pain. Contrary to conventional beliefs about immune cells being passive observers in the nervous system, they are now recognized to actively participate in communication with neurons, contributing to the maintenance of disease states, including neuropathic pain. Plastic changes at the site of nerve injury, dorsal root ganglia (DRG), and spinal cord dorsal horn have been documented.

Endothelial damage and elevated neuronal activity together have been identified as factors contributing to the peripheral recruitment of monocytes/macrophages and the activation of microglia. This activation leads to the release of mediators that cause neuronal sensitization, establishing a positive feedback loop that sustains chronic pain. In cases of peripheral nerve injury in cancer, endothelial activation triggers the recruitment of monocytes/macrophages expressing CX3C chemokine receptor 1 (CX3CR1). These cells alert nociceptive neurons by releasing reactive oxygen species (ROS), activating transient receptor potential ankyrin 1 (TRPA1) channels, and inducing a pain response. As a result of peripheral nerve injury, neuro-immune crosstalk occurs in the dorsal root ganglion (DRG), driving release of extracellular vesicles, which are then engulfed by nearby macrophages. A growing body of evidence over the last two decades has revealed signaling pathways that facilitate communication between neurons and microglia within the dorsal horn of the spinal cord. These pathways include P2X4/BDNF/GABAA, P2X7/Cathepsin S/Fractalkine/CX3CR1, and CSF-1/CSF-1R/DAP12, each playing a role in the mechanisms of communication. Understanding the modalities of communication between neurons and immune cells at different locations along the pain pathway in neuropathic pain presents opportunities for the development of novel approaches in the treatment of chronic pain. (Malcangio, 2019). In gastric cancer, both cancer cells and nerve terminals synthesized and released nervous system-associated factors to mediate neuro-immune interactions, actively contribute to the regulation of cell proliferation, angiogenesis, metastasis, and recurrence. Notably, cancer cells invades nerve fibers and induce the outgrowth and branching of neural cells. This phenomenon enables the infiltration of nerve fibers into the TME, ultimately promoting the progression of gastric cancer (Wang et al., 2020).

Natural immune mediators, such as cytokines and hormones, which also induce acute phase responses (APR), are implicated in conferring resistance against tumor development. Additionally complex neuroimmune mechanisms and variations contribute to the host's defense against cancer, shedding light on the role of the adaptive immune system in individuals with tumors. Defence mechanisms are further enhanced by specifically designed vaccines and cytokines. Leveraging hormonal stimulation of immune mechanisms, in conjunction with established immunostimulants, holds promise for the advancement of combination immunotherapy in cancer treatment (Berczi et al., 1998).

Behavioral comorbidities, such as depression, anxiety, fatigue, cognitive disturbances, and neuropathic pain, are widespread among cancer patients and survivors, significantly diminishing their quality of life across various types of cancer. Changes in neuroimmune and neuroinflammatory interactions have been implicated in altering affective-like or cognitive behaviors in rodent models, specifically, changes in microglia, neuroinflammation, and immune trafficking to the brain are observed in tumor models received chemotherapy and/or radiation (Santos & Pyter, 2018).

The neuro-immune axis is crucial for understanding the normal function of the CNS and the pathophysiology of various brain disorders, offering promising prospects as novel therapeutic targets. Microglia, a type of glial cell, is recognized for playing significant roles in brain disorders. Astrocytes, on the other hand, have been observed to exhibit both positive and negative modulation in the progression of multiple CNS disorders. The interaction between reactive astrocytes and the neuro-immune axis holds particular importance in the context of brain tumors. Studies on CNS disorders have unveiled that reactive astrocytes serve as a key component of the neuro-immune axis, actively modulating both the innate and adaptive immune systems (Priego & Valiente 2019).

The PNS allows animals to sense and respond to changes in tissue homeostasis and repair. Studies indicate that PNS play a role in modulating stem cell behaviors and immune responses under steady-state conditions, as well as in response to injury and stress. This modulation is particularly observed in the skin and the hematopoietic system, showcasing neuro-stem cell and neuro-immune cell interactions that connect tissue biology and local immunity with systemic bodily changes to meet varying demands (Scott-Solomon & Hsu 2022). The use of intravital and ex vivo two-photon laser scanning microscopy (2P-LSM) has unveiled novel pathways that facilitate communication between immune cells, stromal cells, cancer cells, and proteins in the CNS parenchyma and peripheral vasculature. The combination of intravital 2P-LSM and microfluidic devices, coupled with mathematical modeling, has provided new insights into the intravascular behaviour of circulating immune cells (Dorand et al., 2019).

Growing evidences indicate links between social support, stress, emotional state, autonomic nervous systems (ANS), and immune and endocrine function. Certain mechanisms, such as NK cells and cortisol, are considered mediators in the relationship between stress, social support, and cancer progression, supported by solid basic and clinical evidence (Spiegel & Sephton 2001).

Semaphorin 4D (Sema4D) or CD100, has been implicated in pathophysiological roles in both the immune and nervous systems. Sema4D-mediated cell signaling, especially interaction between Sema4D and its high-affinity receptor, Plexin-B1, plays a significant role in both physiological and pathological processes in human disease, particularly within the TME during tumor angiogenesis and metastasis (Wu et al., 2016).

Patients undergoing cancer treatment often face a significant symptom burden, and addressing disease-related symptoms may enhance overall tolerance to treatment. Chronic emotional and psychosocial stressors can pose a risk of neuroimmune dysfunction in BC survivors. With rural survivors facing more challenges than urban BC survivors highlighted a notable discrepancy. Consequently, there is a clear need to investigate the impact of rural-urban residence as a factor in comprehending the connections between neuroimmune function (such as sAA and IL-6) and perceived health status, especially regarding social functioning in women with breast cancer (BC) (Hulett et al., 2021).

Tumor-induced inflammation is hypothesized to contribute to disease-related symptoms. In a murine model of human papillomavirus (HPV)-related head and neck cancer, a clear relationship between cytokine levels and mild behavioral changes was observed. The tumor progression led to elevated expression of inflammatory cytokines (IL-6, IL-1β, TNF-α) in the liver and increased IL-1β expression in the brain. Tumor tissue itself exhibited high inflammatory cytokine expression compared to healthy tissue. A low dose of lipopolysaccharide (LPS, 0.1mg/kg) treatment caused elevated serum IL-6, deficits in nest building, tail suspension, and locomotor activity in tumor-bearing mice without treatment induced weight loss, anorexia, or anhedonia. Minocycline, an anti-inflammatory agent, attenuated tumor-induced inflammation and burrowing deficits, established an inflammatory-like mechanism for behavioral alterations associated with tumor growth though the inflammatory state and behavioral changes.

5.11 TARGETING THE ADRENERGIC RECEPTORS TO MANAGE CANCER

Targeting specific components of this intricate network may offer novel therapeutic strategies for cancer treatment. Modulating the nervous system's influence on immune responses can potentially enhance antitumor immunity and improve the efficacy of immunotherapies. Adrenergic inhibitors are employed in the treatment of psychological and physiological stress (Pathak & Mrabeti, 2021). Beta-adrenergic receptor blocker usage in cancer treatment is associated with prolonged survival in melanoma, pancreatic, ovarian, and breast cancers. The beta-2-adrenergic receptor antagonists exhibit greater responsiveness alongside immunotherapy. (Kokolus et al., 2017; Yap et al., 2018). The ablation of sympathetic nerves or use of beta-blockers decreases stress signaling and tumor growth (Scheff et al., 2021).

5.12 CONCLUSION

The intricate interplay between the nervous and immune systems during cancer reveals a dynamic relationship that significantly influences tumor development and progression. Furthermore, elevated levels of intertumoral nerve density associates with poor prognosis and high chances of recurrence across different solid tumors (Silverman et al., 2021). From the neuroimmune crosstalk promoting tumor growth to the modulation of immune responses by neural components, these interactions are pivotal in shaping the TME and impacting clinical outcomes. Understanding the neuroimmune crosstalk provides valuable insights into cancer biology and opens avenues for innovative therapeutic strategies. Targeting the nervous system not only enhances antitumor immunity but also holds promise in alleviating cancer-associated neurological symptoms, such as chronic pain, fatigue, and cognitive impairment. The recent focus on CNS and peripheral nervous system control of tumorigenesis and metastasis has unveiled novel strategies for cancer targeting and treatment.

Axonogenesis and neurogenesis signify homeostasis and energetic metabolism, while PNI is characterized by increased proliferation and decreased apoptosis in tumor cells and high nerve density in tumors. Ultimately, carcinoma cells undergo transdifferentiation into a neuronal profile, representing the final stage in neuroepithelial interactions. Moreover, the role of neuroimmune interactions extends beyond tumor growth, influencing responses to immunotherapy and affecting the prognosis of various cancers. The integration of neural regulation into cancer treatment approaches becomes paramount for comprehensive and effective therapeutic strategies (Ayala & Chen, 2018; Schmitd et al., 2023).

As research continues to unravel the complexities of neuroimmune crosstalk in cancer, the clinical implications become increasingly evident. Tailoring interventions that consider the bidirectional communication between the nervous and immune systems may pave the way for more personalized and targeted cancer therapies, ultimately improving patient outcomes and quality of life.

ABBREVIATION

central nervous system (CNS)
myeloid-derived suppressor cells (MDSCs)
peripheral nervous system (PNS)
sympathetic nervous system (SNS)
tumor microenvironment (TME)
tumor necrosis factor alpha (TNF-α)

programmed cell death-1 (PD1)
extracellular vesicles (EVs)
micro RNA (miRNA)
natural killer (NK) cell
perineural invasion (PNI)
nerve growth factor family (NGF)
Glial cell-derived neurotrophic factor (GDNF)
midkine (MDK) family
semaphorin 3D (SEMA3D)
Cancer associated Fibroblasts (CAFs)
nectin cell adhesion molecule 1 (NECTIN1)
tumor-associated macrophages (TAMs)

REFERENCES

Abramsky O, Wertman E, Reches A, Brenner T, Ovadia H. Effect of hypothalamic lesions on experimental autoimmune diseases in rats. *Ann N Y Acad Sci.* 1987;496:360–5. doi: 10.1111/j.1749-6632.1987.tb35788.x. PMID: 2440369.

Ader R. Historical Perspectives on Psychoneuroimmunology. *Psychoneuroimmunology, stress and infection*, CRC Press, Boca Raton, 1995 (pp. 1–21)

Airaksinen MS, Saarma M. The GDNF family: signalling, biological functions and therapeutic value. *Nat Rev Neurosci.* 2002 May;3(5):383–94. doi: 10.1038/nrn812. PMID: 11988777.

Ayala GE, Chen D. Innervating prostate cancer. *N Engl J Med.* 2018;378(7):675–7. doi: 10.1056/NEJMcibr1714003. PMID: 29443663.

Ayala GE, Dai H, Powell M, Li R, Ding Y, Wheeler TM, Shine D, Kadmon D, Thompson T, Miles BJ, Ittmann MM, Rowley D. Cancer-related axonogenesis and neurogenesis in prostate cancer. *Clin Cancer Res.* 2008 Dec 1;14(23):7593–603. doi: 10.1158/1078-0432.CCR-08-1164. PMID: 19047084.

Ayala GE, Wheeler TM, Shine HD, Schmelz M, Frolov A, Chakraborty S, Rowley D. In vitro dorsal root ganglia and human prostate cell line interaction: Redefining perineural invasion in prostate cancer. *Prostate.* 2001 Nov 1;49(3):213–23. doi: 10.1002/pros.1137. PMID: 11746267.

Bapat AA, Hostetter G, Von Hoff DD, Han H. Perineural invasion and associated pain in pancreatic cancer. *Nat Rev Cancer.* 2011 Sep 23;11(10):695–707. doi: 10.1038/nrc3131. PMID: 21941281.

Barni S, Lissoni P, Crispino S, Cattaneo G, Rovelli F, Fumagalli G, Tancini G. Neuroimmunomodulation in cancer patients: Correlations between melatonin and beta-endorphin blood levels and T helper/suppressor ratio. *Int J Biol Markers* 1988;3(2):82–6. doi: 10.1177/172460088800300202. PMID: 2977397.

Berczi I, Chow DA, Baral E, Nagy E. Neuroimmunoregulation and cancer (review). *Int J Oncol.* 1998 Nov;13(5):1049–60.

Cole SW, Nagaraja AS, Lutgendorf SK, Green PA, Sood AK. Sympathetic nervous system regulation of the tumor microenvironment. *Nat Rev Cancer.* 2015 Sep;15(9):563–72. doi: 10.1038/nrc3978. PMID: 26299593; PMCID: PMC4828959.

Cruveilheir J *Maladies Des Nerfs Anatomie Pathologique Du Corps Humain.* 2nd ed. Paris, France: JB Bailliere 1835.

Dantzer R. Neuroimmune interactions: From the brain to the immune system and vice versa. *Physiol Rev.* 2018 Jan 1;98(1):477–504. doi: 10.1152/physrev.00039.2016. PMID: 29351513; PMCID: PMC5866360.

Dantzer R, Cohen S, Russo S J, Dinan T G. Resilience and immunity. *Brain Behav Immun.* 2018;74:28–42. doi: 10.1016/j.bbi.2018.08.010. PMID: 30102966.

Deborde S, Omelchenko T, Lyubchik A, Zhou Y, He S, McNamara WF, Chernichenko N, Lee SY, Barajas F, Chen CH, Bakst RL, Vakiani E, He S, Hall A, Wong RJ. Schwann cells induce cancer cell dispersion and invasion. *J Clin Invest.* 2016 Apr 1;126(4):1538–54. doi: 10.1172/JCI82658. Epub 2016 Mar 21. PMID: 26999607; PMCID: PMC4811155.

Dorand RD, Benson BL, Huang LF, Petrosiute A, Huang AY. Insights from dynamic neuroimmune imaging on murine immune responses to CNS damage. *Front Neurosci.* 2019 Jul 17;13:737.

Faith RE, Liang HJ, Plotnikoff NP, Murgo AJ, Nimeh NF. Neuroimmunomodulation with enkephalins: In vitro enhancement of natural killer cell activity in peripheral blood lymphocytes from cancer patients. *Nat Immun Cell Growth Regulation.* 1987;6(2):88–98. PMID: 3600678.

Guo K, Ma Q, Li J, Wang Z, Shan T, Li W, Xu Q, Xie K. Interaction of the sympathetic nerve with pancreatic cancer cells promotes perineural invasion through the activation of STAT3 signaling. *Mol Cancer Ther.* 2013 Mar;12(3):264–73. doi: 10.1158/1535-7163.MCT-12-0809. Epub 2013 Jan 3. PMID: 23288783.

Hodo TW, de Aquino MTP, Shimamoto A, Shanker A. Critical Neurotransmitters in the Neuroimmune Network. *Front Immunol.* 2020 Aug 21;11:1869.

Horvathova L, Tillinger A, Sivakova I, Mikova L, Mravec B, Bucova M. Chemical sympathectomy increases neutrophil-to-lymphocyte ratio in tumor-bearing rats but does not influence cancer progression. *J Neuroimmunol.* 2015 Jan 15;278:255–61. doi: 10.1016/j.jneuroim.2014.11.014. Epub 2014 Nov 20. PMID: 25468774.

Hulett JM, Abshire DA, Armer JM, Millspaugh R, Millspaugh J. Rural-Urban differences in neuroimmune biomarkers and health status among women living with breast cancer. *Cancer Nurs.* 2021 Jul-Aug 01;44(4):323–332.

Jean Wrobel L, Bod L, Lengagne R, Kato M, Prévost-Blondel A, Le Gal FA. Propranolol induces a favourable shift of anti-tumor immunity in a murine spontaneous model of melanoma. *Oncotarget.* 2016 Nov 22;7(47):77825–77837. doi: 10.18632/oncotarget.12833. PMID: 27788481; PMCID: PMC5363624.

Jurcak NR, Rucki AA, Muth S, Thompson E, Sharma R, Ding D, Zhu Q, Eshleman JR, Anders RA, Jaffee EM, Fujiwara K, Zheng L. Axon guidance molecules promote perineural invasion and metastasis of orthotopic pancreatic tumors in mice. *Gastroenterology.* 2019 Sep;157(3):838–850.e6. doi: 10.1053/j.gastro.2019.05.065. Epub 2019 Jun 1. PMID: 31163177; PMCID: PMC6707836.

Kokolus KM, Zhang Y, Sivik JM, Schmeck C, Zhu J, Repasky EA, Drabick JJ, Schell TD. Beta blocker use correlates with better overall survival in metastatic melanoma patients and improves the efficacy of immunotherapies in mice. *Oncoimmunology.* 2017 Dec 21;7(3):e1405205. doi: 10.1080/2162402X.2017.1405205. PMID: 29399407; PMCID: PMC5790362.

Li J, Kang R, Tang D. Cellular and molecular mechanisms of perineural invasion of pancreatic ductal adenocarcinoma. *Cancer Commun (Lond).* 2021 Aug;41(8):642–660. doi: 10.1002/cac2.12188. Epub 2021 Jul 15. PMID: 34264020; PMCID: PMC8360640.

Liebig C, Ayala G, Wilks JA, Berger DH, Albo D. Perineural invasion in cancer: a review of the literature. *Cancer.* 2009 Aug 1;115(15):3379–91. doi: 10.1002/cncr.24396. PMID: 19484787.

Magnon C, Hondermarck H. The neural addiction of cancer. *Nat Rev Cancer.* 2023 May;23(5): 317–334. doi: 10.1038/s41568-023-00556-8. Epub 2023 Apr 11. PMID: 37041409.

Malcangio M. Role of the immune system in neuropathic pain. *Scand J Pain.* 2019 Dec 18;20(1):33–37.

Mauffrey P, Tchitchek N, Barroca V, Bemelmans AP, Firlej V, Allory Y, Roméo PH, Magnon C. Progenitors from the central nervous system drive neurogenesis in cancer. *Nature.* 2019 May;569(7758):672–678. doi: 10.1038/s41586-019-1219-y. Epub 2019 May 15. Erratum in: Nature. 2020 Jan;577(7792):E10. PMID: 31092925.

Miknyoczki SJ, Lang D, Huang L, Klein-Szanto AJ, Dionne CA, Ruggeri BA. Neurotrophins and Trk receptors in human pancreatic ductal adenocarcinoma: expression patterns and effects on in vitro invasive behaviour. *Int J Cancer.* 1999 May 5;81(3):417–427.

Moran A, Takahashi H, Dragomir MP, Lindemann A, Gleber-Netto FO, Pickering CR, Anfossi S, Osman AA, Cai Y, Wang R, Knutsen E, Shimizu M, Ivan C, Rao X, Wang J, Silverman DA, Tam S, Zhao M, Caulin C, Zinger A, Tasciotti E, Dougherty PM, El-Naggar A, Calin GA, Myers JN. Loss of p53 drives neuron reprogramming in head and neck cancer. *Nature* 2020; 578(7795):449–454. doi: 10.1038/s41586-020-1996-3. PMID: 32051587.

Na'ara S, Amit M, Gil Z. L1CAM induces perineural invasion of pancreas cancer cells by upregulation of metalloproteinase expression. *Oncogene*. 2019;38(4):596–608.

Neumann E. Secondare cancroid infiltration des nervus mentalis bei einem. *Arch Pathol Anat*. 1862;24:201–201.

Oertel H. Innervation and tumor growth: A preliminary report. *Can Med Assoc J*. 1928 Feb;18(2):135–9. PMID: 20316697; PMCID: PMC1709538.

Olar A, He D, Florentin D, Ding Y, Ayala G. Biologic correlates and significance of axonogenesis in prostate cancer. *Hum Pathol*. 2014 Jul;45(7):1358–64. doi: 10.1016/j.humpath.2014.02.009. Epub 2014 Feb 28. PMID: 24767770; PMCID: PMC4492299.

Pathak A, Mrabeti S. β-Blockade for patients with hypertension, ischemic heart disease or heart failure: Where are we now? *Vasc Health Risk Manag*. 2021 Jun 8;17:337–348. doi: 10.2147/VHRM.S285907. PMID: 34135591; PMCID: PMC8197620.

Priego N, Valiente M. The potential of astrocytes as immune modulators in brain tumors. *Front Immunol*. 2019 Jun 11;10:1314.

Reavis HD, Chen HI, Drapkin R. Tumor innervation: Cancer has some nerve. *Trends Cancer*. 2020 Dec;6(12):1059–1067. doi: 10.1016/j.trecan.2020.07.005. Epub 2020 Aug 14. PMID: 32807693; PMCID: PMC7688507.

Ribatti D. The concept of immune surveillance against tumors: The first theories. *Oncotarget*. 2017 Jan 24;8(4):7175–7180. doi: 10.18632/oncotarget.12739. PMID: 27764780; PMCID: PMC5351698.

Santos JC, Pyter LM. Neuroimmunology of behavioural comorbidities associated with cancer and cancer treatments. *Front Immunol*. 2018 Jun 7;9:1195.

Scanlon CS, Banerjee R, Inglehart RC, Liu M, Russo N, Hariharan A, van Tubergen EA, Corson SL, Asangani IA, Mistretta CM, Chinnaiyan AM, D'Silva NJ. Galanin modulates the neural niche to favour perineural invasion in head and neck cancer. *Nat Commun*. 2015 Apr 28;6:6885. doi: 10.1038/ncomms7885. PMID: 25917569; PMCID: PMC4476386.

Scheff NN, Saloman JL. Neuroimmunology of cancer and associated symptomology. *Immunol Cell Biol*. 2021 Oct;99(9):949–61. doi: 10.1111/imcb.12496. Epub 2021 Sep 12. PMID: 34355434; PMCID: PMC9250294.

Schmitd LB, Perez-Pacheco C, D'Silva NJ. Nerve density in cancer: Less is better. *FASEB Bioadv*. 2021 Jul 11;3(10):773–86. doi: 10.1096/fba.2021-00046. PMID: 34632313; PMCID: PMC8493966.

Schmitd LB, Perez-Pacheco C, Furgal A, Bellile EL, Liu M, Fattah A, Gonzalez-Maldonado L, Unsworth SP, Wong SY, Rozek LS, Rao A, Wolf GT, Taylor JMG, Casper K, Mierzwa M, D'Silva NJ. Increased nerve density adversely affects outcome in oral cancer. *Clin Cancer Res*. 2023;29(13):2501–12. doi: 10.1158/1078-0432.CCR-22-3496. PMID: 37039710.

Scott-Solomon E, Hsu YC. Neurobiology, stem cell biology, and immunology: An emerging triad for understanding tissue homeostasis and repair. *Annu Rev Cell Dev Biol*. 2022 Oct 6;38:419–446.

Servick K. War of nerves. *Science*. 2019 Sep 13;365(6458):1071–1073. doi: 10.1126/science.365.6458.1071. Epub 2019 Sep 12. PMID: 31515368.

Sharma D, Farrar JD. Adrenergic regulation of immune cell function and inflammation. *Semin Immunopathol*. 2020 Dec;42(6):709–17. doi: 10.1007/s00281-020-00829-6. Epub 2020 Nov 20. PMID: 33219396; PMCID: PMC7678770.

Silverman DA, Martinez VK, Dougherty PM, Myers JN, Calin GA, Amit M. Cancer-associated neurogenesis and Nerve-cancer cross-talk. *Cancer Res.* 2021 Mar 15;81(6):1431–40. doi: 10.1158/0008-5472.CAN-20-2793. Epub 2020 Dec 17. PMID: 33334813; PMCID: PMC7969424.

Spiegel D, Sephton SE. Psychoneuroimmune and endocrine pathways in cancer: Effects of stress and support. *Semin Clin Neuropsychiatry.* 2001 Oct;6(4):252–65.

Taams LS. Neuroimmune interactions: How the nervous and immune systems influence each other. *Clin Exp Immunol.* 2019 Sep;197(3):276–277.

Thaker PH, Han LY, Kamat AA, Arevalo JM, Takahashi R, Lu C, Jennings NB, Armaiz-Pena G, Bankson JA, Ravoori M, Merritt WM, Lin YG, Mangala LS, Kim TJ, Coleman RL, Landen CN, Li Y, Felix E, Sanguino AM, Newman RA, Lloyd M, Gershenson DM, Kundra V, Lopez-Berestein G, Lutgendorf SK, Cole SW, Sood AK. Chronic stress promotes tumor growth and angiogenesis in a mouse model of ovarian carcinoma. *Nat Med.* 2006 Aug;12(8):939–44. doi: 10.1038/nm1447. Epub 2006 Jul 23. Erratum in: Nat Med. 2021 Dec;27(12):2246. PMID: 16862152.

Wang K, Zhao XH, Liu J, Zhang R, Li JP. Nervous system and gastric cancer. *Biochim Biophys Acta Rev Cancer.* 2020 Jan;1873(1):188313.

Wu M, Li J, Gao Q, Ye F. The role of Sema4D/CD100 as a therapeutic target for tumor microenvironments and for autoimmune, neuroimmune and bone diseases. *Expert Opin Ther Targets.* 2016 Jul;20(7):885–901.

Xu Q, Wang Z, Chen X, Duan W, Lei J, Zong L, Li X, Sheng L, Ma J, Han L, Li W, Zhang L, Guo K, Ma Z, Wu Z, Wu E, Ma Q. Stromal-derived factor-1α/CXCL12-CXCR4 chemotactic pathway promotes perineural invasion in pancreatic cancer. *Oncotarget.* 2015 Mar 10;6(7): 4717–32. doi: 10.18632/oncotarget.3069. PMID: 25605248; PMCID: PMC4467110.

Yamada M, Hirabayashi K, Kawanishi A, Hadano A, Takanashi Y, Izumi H, Kawaguchi Y, Mine T, Nakamura N, Nakagohri T. Nectin-1 expression in cancer-associated fibroblasts is a predictor of poor prognosis for pancreatic ductal adenocarcinoma. *Surg Today.* 2018 May;48(5):510–516. doi: 10.1007/s00595-017-1618-3. Epub 2017 Dec 18. PMID: 29256146.

Yao J, Li WY, Li SG, Feng XS, Gao SG. Midkine promotes perineural invasion in human pancreatic cancer. *World J Gastroenterol.* 2014 Mar 21;20(11):3018–24. doi: 10.3748/wjg.v20.i11.3018. PMID: 24659893; PMCID: PMC3960408.

Yap A, Lopez-Olivo MA, Dubowitz J, Pratt G, Hiller J, Gottumukkala V, Sloan E, Riedel B, Schier R. Effect of beta-blockers on cancer recurrence and survival: A meta-analysis of epidemiological and perioperative studies. *Br J Anaesth.* 2018 Jul;121(1):45–57. doi: 10.1016/j.bja.2018.03.024. Epub 2018 May 3. PMID: 29935594.

Young HH. On the presence of nerves in tumors and of other structures in them as revealed by a modification of Ehrlich's method of "vital staining" with methylene blue. *J Exp Med.* 1897 Jan 1;2(1):1–12. doi: 10.1084/jem.2.1.1. PMID: 19866822; PMCID: PMC 2117917.

Zeng L, Guo Y, Liang J, Chen S, Peng P, Zhang Q, Su H, Chen Y, Huang K. Perineural invasion and TAMs in pancreatic ductal adenocarcinomas: Review of the original pathology reports using immunohistochemical enhancement and relationships with clinicopathological features. *J Cancer.* 2014 Oct 22;5(9):754–60. doi: 10.7150/jca.10238. PMID: 25368675; PMCID: PMC4216799.

Zeng Q, Cheng Y, Zhu Q, Yu Z, Wu X, Huang K, Zhou M, Han S, Zhang Q. The relationship between overexpression of glial cell-derived neurotrophic factor and its RET receptor with progression and prognosis of human pancreatic cancer. *J Int Med Res.* 2008 Jul-Aug;36(4):656–64. doi: 10.1177/147323000803600406. PMID: 18652760.

6 Drosophila Models of Neurodegenerative Disorders

Ashwin Kumar Jainarayanan
University of Oxford, Oxford, UK

Arpit Kumar Pradhan
University of Michigan, Ann Arbor, MI, USA

6.1 INTRODUCTION

Neuronal pathology affects the neuronal integrity disturbing the synaptic homeostasis, leading to more diverse pathological symptoms. Loss of synapse has often been attributed to cognitive decline, memory impairment, and behavioral abnormality. In the previous chapters, we discussed the cross-communication between the neuronal and immune systems as well as discussed how disturbances on a molecular level can span on a systemic scale affecting the well-being of the individuals. In this chapter, we will understand the fly model of neurodegeneration and how these small organisms can help understand the molecular changes that affect the pathophysiology of neurodegeneration.

6.2 PARKINSON'S DISEASE

Parkinson's Disease (PD) is a prevalent and debilitating movement disorder, ranking as the second most common neurodegenerative condition (Pandey and Nichols 2011). It is characterized by a range of motor symptoms, including muscle rigidity, resting tremors, bradykinesia (slowness of movement), and postural instability. These clinical features collectively result from a complex interplay of biological factors. At the core of PD pathology is the depletion of dopamine in the brain due to the progressive degeneration of dopaminergic neurons located in a region called the substantia nigra (Feany and Bender 2000). Dopamine is a neurotransmitter crucial for coordinating smooth and controlled movements, and its deficiency leads to the motor symptoms observed in PD.

One of the distinctive hallmarks of PD is the formation of Lewy bodies, which are abnormal protein aggregates found within the cytoplasm of affected neurons. These aggregates primarily consist of a protein called α-Synuclein (α-Syn) and ubiquitin (Spillantini et al. 1997). While the exact role of Lewy bodies in PD is still under

 DOI: 10.1201/9781003295471-6

investigation, their presence is considered a pathological signature of the disease. While the majority of PD cases do not have a known cause, postmortem studies of sporadic cases have provided critical insights. These studies have revealed evidence of defective mitochondrial function and oxidative damage specifically in dopaminergic neurons (Schapira et al. 1990).

Further supporting the role of mitochondrial dysfunction and oxidative stress in PD are studies involving environmental toxins like 1-methyl-4-phenyl-1,2,3,6-tetrahydropyridine (MPTP) and rotenone (Langston et al. 1983). Genetic factors also play a significant role in PD. There are rare familial forms of the disease associated with specific genetic mutations, and researchers have identified several genetic loci (PARK1–11) linked to familial PD (Krüger 2004). Among the known familial PD (FPD) genes are α-Syn and Parkin.

Recent studies have unveiled connections between these FPD genes and mitochondrial dysfunction and oxidative stress, highlighting their involvement in the disease process. Additionally, FPD genes like DJ-1, PINK1, and LRRK2 have been identified, although their precise roles in PD pathogenesis remain somewhat undefined (Martin, Dawson, and Dawson 2011). Mouse models engineered to carry mutations in genes such as α-Syn, Parkin, DJ-1, and Pink1 have provided valuable insights into the genetics of PD. However, these models often fall short of fully replicating one of PD's cardinal features, which is the selective degeneration of dopaminergic neurons (Crabtree and Zhang 2012).

To bridge this gap, researchers are exploring combinations of genetic modifications and alterations in the aging process to better mimic the selective dopaminergic neuron degeneration seen in PD. Understanding these complex mechanisms is crucial for unraveling the pathogenesis of PD and ultimately developing more effective therapeutic strategies for this challenging neurodegenerative disorder.

Drosophila melanogaster, commonly known as the fruit fly, has emerged as a powerful model organism for investigating PD due to its genetic tractability, relatively short lifespan, conservation of cellular processes, and ability to replicate key features of the disease (Pandey and Nichols 2011). The shorter lifespan of Drosophila is advantageous for studying the age-related aspects of PD, enabling the observation of disease progression within a manageable timeframe. The evolutionary conservation of cellular processes between fruit flies and mammals makes Drosophila an ideal model for investigating PD-related mechanisms, despite its simplified system.

Furthermore, the fruit fly's dopaminergic system closely resembles that of mammals, providing insights into how dysfunction in dopaminergic neurons impacts motor behavior and other PD-related phenotypes (Feany and Bender 2000). Drosophila models allow for the direct observation of neurodegeneration in dopaminergic neurons. Live imaging techniques can be employed to monitor the health and survival of these neurons over time, providing insights into the progression of neurodegeneration and potential interventions to halt or slow it.

Researchers have successfully generated Drosophila models expressing human α-Syn or mutant forms of genes associated with FPD, faithfully replicating key PD features such as α-Syn aggregation, dopaminergic neuron loss, locomotor deficits, and reduced lifespan (Warrick et al. 1999). This faithful representation enables scientists to investigate underlying molecular pathways involved in α-Syn toxicity, mitochondrial dysfunction, oxidative stress responses, and protein clearance mechanisms, contributing to a deeper understanding of PD pathogenesis (Auluck et al. 2002).

Drosophila's behavior is highly amenable to analysis, making it an ideal model for studying motor and cognitive deficits associated with PD. Researchers can conduct a wide range of behavioral assays to assess locomotor activity, coordination, memory, and learning in fruit fly models of PD. These assays provide quantitative data on disease-related phenotypes. Further, Drosophila models can be exposed to environmental toxins known to contribute to PD, such as MPTP and rotenone. This allows researchers to investigate how these toxins induce mitochondrial dysfunction and oxidative stress, replicating key environmental factors associated with the disease.

Furthermore, Drosophila-based studies have significantly contributed to the understanding of autophagy and lysosomal pathways, crucial for PD-related protein degradation. Genetic screens in flies have identified potential modifiers of PD pathology, opening new avenues for drug discovery. Compounds that rescue PD-related phenotypes in fruit fly models offer promise for further testing in mammalian systems and, potentially, in clinical trials, representing a beacon of hope for the development of novel PD treatments (Nixon, Yang, and Lee 2008). Drosophila models are particularly well-suited for high-throughput screening of potential therapeutic interventions. Researchers can test a wide range of compounds, including drugs, gene therapies, and other interventions, to assess their effectiveness in mitigating PD-related phenotypes cost-effectively and efficiently. Compounds that show promise in Drosophila models can then be further evaluated in mammalian systems and potentially progress to clinical trials (Bilen and Bonini 2005).

In essence, the combination of genetic and environmental factors contributing to PD pathogenesis is a challenging puzzle, but the powerful Drosophila model organism has provided invaluable insights and potential therapeutic targets, bringing us closer to unraveling the mysteries of Parkinson's Disease and advancing the quest for effective treatments.

6.3 POLYGLUTAMINE (POLY Q) RELATED DISORDERS

Drosophila melanogaster, commonly known as the fruit fly, has established itself as a valuable and versatile model organism for the investigation of polyglutamine (Poly Q) related disorders. This group of neurodegenerative diseases encompasses conditions characterized by the abnormal expansion of CAG trinucleotide repeats within specific disease-associated proteins, resulting in the elongation of a polyglutamine stretch. Notable among these disorders are Huntington's disease (HD) and various spinocerebellar ataxias (SCAs), collectively referred to as Poly Q diseases.

The choice of Drosophila as a model for studying Poly Q diseases is grounded in several compelling advantages. First, its genetic tractability allows for the precise manipulation of the expression of disease-associated Poly Q-expanded proteins, enabling researchers to dissect the intricate pathogenic mechanisms that underlie these devastating conditions. This genetic manipulability is particularly crucial for understanding the effects of varying Poly Q repeat lengths and their impact on disease progression. Additionally, Drosophila's relatively short lifespan is advantageous for investigating the age-related aspects of Poly Q diseases. This shortened lifespan allows for the observation of disease progression and its effects on lifespan within a manageable timeframe, offering insights into how these disorders manifest and evolve.

Transgenic Drosophila models have been instrumental in recapitulating key features of Poly Q diseases, faithfully reproducing disease-specific molecular events and pathological hallmarks. These models have yielded critical insights into the cellular and molecular processes involved in the pathogenesis of Poly Q diseases (Lewis and Smith 2016; Jackson et al. 1998). Notably, studies in Drosophila Poly Q models have illuminated how expanded Poly Q repeats can lead to protein misfolding and aggregation, disrupting normal cellular functions. These aggregated proteins often accumulate within specific neuronal populations, contributing to selective neurodegeneration. Furthermore, Drosophila-based investigations have shed light on the role of protein quality control systems, including chaperones and autophagy, in mitigating the toxic effects of Poly Q-expanded proteins (Warrick et al. 1999; Jaiswal et al. 2009).

Drosophila Poly Q models have also been instrumental in the identification of genetic modifiers and prospective therapeutic targets for Poly Q diseases. Through genetic screens and the manipulation of candidate genes, researchers have unearthed factors capable of either enhancing or suppressing disease phenotypes. These findings unveil promising pathways for the development of novel therapeutic interventions (Pandey et al. 2007; Fernandez-Funez et al. 2000). Behavioral assays conducted within Drosophila Poly Q models have demonstrated motor deficits and altered neuronal functions that bear a striking resemblance to human disease symptoms. These assays offer a high-throughput platform for drug screening, enabling the identification of compounds with the potential to rescue or ameliorate disease-associated phenotypes (Wang et al., 2008; Landles et al., 2010). Additionally, the Drosophila model has been used to study the circadian dysfunction in Huntington's disorder as well as Poly Q disorders. Using the Drosophila model system, we had previously shown that heat shock protein 40 (HSP40) overexpression delays and partly rescues the circadian dysfunction in a Huntington's model (Prakash, Pradhan, and Sheeba 2022).

In essence, Drosophila melanogaster stands as an indispensable ally in advancing our comprehension of the intricacies of Poly Q related disorders, from unraveling molecular intricacies to identifying genetic modifiers and potential therapeutics. This research underscores the pivotal role of Drosophila as a versatile and powerful model organism in the relentless pursuit of combatting the devastating impact of neurodegenerative disorders like HD and SCAs.

6.4 AMYOTROPHIC LATERAL SCLEROSIS (ALS)

Amyotrophic Lateral Sclerosis (ALS), a devastating and currently incurable neurodegenerative disorder, has benefited significantly from research conducted in Drosophila melanogaster, commonly known as the fruit fly. Drosophila serves as a powerful and indispensable model organism in the study of ALS, primarily due to its genetic tractability and rapid lifecycle, which together enable comprehensive mechanistic investigations into the pathological processes underlying this devastating disease.

One of the pivotal contributions of Drosophila ALS models is in the study of ALS-associated proteins, such as mutant forms of superoxide dismutase 1 (SOD1). Transgenic Drosophila expressing these mutant ALS-associated proteins have provided crucial insights into the disease. ALS-linked SOD1 mutations within Drosophila result in protein misfolding, aggregation, and, most significantly, motor neuron degeneration a striking recapitulation of the key characteristics of ALS observed in humans (Ratnaparkhi et al. 2008). These faithful representations of human ALS pathology within the Drosophila model have proven invaluable in elucidating the underlying molecular and cellular events that drive this neurodegenerative disorder.

Furthermore, investigations within Drosophila ALS models have unveiled perturbations in protein quality control systems, further deepening our understanding of ALS pathogenesis. Chaperones, particularly HSP70 and HSP90, have been shown to mitigate SOD1-induced toxicity, underscoring their protective roles in combating protein misfolding and aggregation (Ratnaparkhi et al. 2008). Additionally, the critical cellular process of autophagy has been implicated in ALS pathogenesis. Disruption of autophagy exacerbates neurodegeneration in Drosophila ALS models, emphasizing the significance of this cellular mechanism in maintaining neuronal health and homeostasis (Simonsen et al. 2008).

Drosophila ALS models have not only helped unravel the role of specific ALS-associated proteins but have also been instrumental in identifying genetic modifiers that influence disease onset and progression. Genetic screens conducted in these models have allowed researchers to manipulate genes like TDP-43, another ALS-associated protein, shedding light on their contributions to neurodegeneration (Godena et al. 2011). Additionally, investigations into the interactions between different ALS-associated proteins within the Drosophila model have revealed synergistic effects that contribute to the complexity of the disease (Bilen and Bonini 2005).

One of the most promising aspects of Drosophila ALS research lies in its potential for drug screening. Compounds targeting various aspects of ALS pathology, including neuroinflammation, oxidative stress, and protein aggregation, have been tested in Drosophila ALS models and have shown promise in ameliorating disease phenotypes (Mulakkal et al. 2014). This approach holds significant potential for the development of novel therapeutic interventions for ALS.

In conclusion, Drosophila melanogaster has emerged as an indispensable ally in the quest to unravel the mysteries of ALS. Through faithful replication of key disease characteristics and mechanistic insights into ALS-associated proteins, protein quality control systems, genetic modifiers, and potential therapeutic compounds, Drosophila ALS models have significantly advanced our understanding of this devastating neurodegenerative disorder. The fruit fly's contribution to ALS research offers hope for the development of effective treatments and, ultimately, a cure for this debilitating disease.

6.5 ALZHEIMER'S DISEASE

Given the intricate molecular mechanism that governs the pathophysiology of Alzheimer's disease (AD), several model systems ranging from cell culture systems to yeast and *C. elegans* have been used to address different research questions. Every model system has its own sets of positives as well as certain limitations. The usage is therefore based on the research questions being addressed by the investigator. For instance, while the rodent model system has been extensively used to study the pathophysiological mechanisms of AD because it can closely reflect the pathological hallmarks of AD, it has certain limitations which include time, cost, as well as handling and care. On the other hand, small invertebrates such as Drosophila are short-lived which allows faster acquisition of data and can be modeled according to the specific research problem. However, they do not closely resemble the pathological hallmarks of AD as in the case of humans (Lu and Vogel 2009).

An added advantage of using Drosophila as a model system is that its genetic as well as functional anatomy has been extensively studied. While it's a small model organism, its implications in behavioral responses, including locomotion learning and memory, are well-established (Lu and Vogel 2009). The most important hallmarks of AD, $A\beta_{42}$, and tau proteins on overexpression in a fly model show critical destruction of neurons similar to that in the cases of humans. It is important to note that based on the spatial overexpression of AD-related peptides in different neuronal subsets, different phenotypic models of Drosophila can be generated and used to study differential aspects of AD pathology. To summarize, the usage of the fly as a model system to study AD pathology provides advantages in terms of no ethical restrictions, short generation as well as lifespan, and easy maintenance and genetic manipulation. The limitations, however, include lesser complexity in terms of higher-order brain function, and difference in resembling the pathological hallmarks in humans.

The aggregation of amyloid beta peptide and hyperphosphorylated tau constitute two primary hallmarks of AD. This primarily contributes to the progressive cognitive decline and memory impairments. Amyloid beta has often been thought to be the notorious element behind the progressive pathophysiological alterations in AD (Amyloid cascade hypothesis) (Reitz 2012; Knobloch et al. 2007; Castellani, Plascencia-Villa, and Perry 2019; Pimplikar 2009). Several studies however have shown that Aβ itself might not be necessary to progress the disease further. Therefore, it is thought that the aggregation of Aβ initiates downstream processes including tau phosphorylation, which further drive the pathology in AD (Moloney et al. 2010). This is evident from the study that suggests that amyloid plaque load in human subjects does not clinically correlate with AD pathology, rather soluble Aβ oligomers and hyperphosphorylated tau correlate with the pathophysiology (Morris, Clark, and Vissel 2014). From the cell culture studies, it was established intermediate size Aβ aggregates (between monomers and fibrils) were toxic (Townsend et al. 2006; Cleary et al. 2005; Gong et al. 2003; Walsh et al. 2002; Lambert et al. 1998). Several studies from the fly models have positively correlated the extent of AD pathology with the toxic aggregates. For instance, using sequence-dependent methods, Luheshi and colleagues examined Aβ aggregation and toxicity (Pawar et al. 2005; Luheshi et al. 2007). Fly models were used to test their neurotoxicity, and reduced longevity and locomotor performance were found to correlate well with the predicted propensity to form amyloid fibrils. The flies expressing $A\beta_{1\text{-}42}$ had memory impairments, which sheds an important insight into oligomeric Aβ being the primary culprit in the AD pathogenesis (Moloney et al. 2010). The Aβ aggregates instigate a series of downstream events, including protein kinase activation by the transient influx of calcium (Moloney et al. 2010).

While the Drosophila model system consists of an APP ortholog dAPP1 which shares certain similarities with the vertebrate APP, the region corresponding to the Aβ peptide does not share extensive similarities (Luo, Tully, and White 1992). Additionally, however, the fly model contains different components of the γ-secretase enzyme (Periz and Fortini 2004). The fly model system therefore does not endogenously produce Aβ peptide. However, excessive expression of β secretase-like protein in the flies leads to dAPP1 cleavage, which leads to Aβ peptide-like fragment (Carmine-Simmen et al. 2009). This Aβ like fragment has been shown to aggregate and cause behavioral abnormalities as in AD (Carmine-Simmen et al. 2009). A triple transgenic model of fly, which expresses human β-secretase (hBACE), human APP (hAPP), and Drosophila γ-secretase presenilin (dPsn), shows phenotypic differences in terms of loss of neurons, photoreceptor cell loss, as well as axonal degeneration (Ye and Fortini 1999; Greeve et al. 2004). hAPP and hBACE co-expression in Drosophila resulted in forming plaques of hAPP glycosylated species (Ye and Fortini 1999; Greeve et al. 2004).

Researchers have successfully created a reliable invertebrate model to study Aβ toxicity. They achieved this by introducing genetically modified flies with gal4-driven constructs carrying human APP and human BACE1. When these genes are expressed in the fly's brain, human APP undergoes cleavage first by the transgenic human BACE1 and then by the fly's γ-secretase, leading to the formation of the Aβ peptide (Greeve et al. 2004; Prüßing, Voigt, and Schulz 2013). The correlation between the extent of intracellular Aβ accumulation and early phenotypes, including locomotor dysfunction and severity, has been well-documented (Prüßing, Voigt, and Schulz 2013). In addition to the Aβ aggregates, fly model of tauopathies, particularly the tau overexpression models, have been created which features phenotypic differences and differences in terms of longevity. The measurement of longevity in flies is a robust test to understand neuronal integrity. In addition to the longevity test, the locomotor tests provide the investigator with a behavioral performance indicator to differentiate the transgenic flies (Prüßing, Voigt, and Schulz 2013). We will discuss the knowledge gained from the fly model of AD in the subsequent part of this chapter.

First, the fly model, which ectopically expresses $A\beta_{1-42}$ in drosophila photoreceptors, showed aggregating as well as amyloidogenic properties (Finelli et al. 2004; Crowther et al. 2005). Similarly, fly models expressing the $A\beta_{1-42}$ showed age-dependent amyloid deposition similar to human subjects (Jeon et al. 2020; Iijima et al. 2008). Second, as discussed earlier, recent studies indicate a major role of soluble Aβ oligomers in AD pathology. In the Drosophila retina, the production of Aβ peptides is associated with age-dependent neurodegeneration in retinal photoreceptor cells (Greeve et al. 2004). Notably, this neurodegeneration occurs before the formation of Aβ plaques, indicating that the toxicity is likely mediated by Aβ oligomers and protofibrils (Jeon et al. 2020). Third, the intraneuronal accumulation of Aβ is thought to be the crucial seeding point leading to extracellular Aβ deposits, further leading to amyloid deposits, neuronal death, and synaptic toxicity. For instance, in the case of the AD patient's brain, oligomeric Aβ is primarily localized in neurons and is associated with lipid membrane (Hong et al. 2014). The metabolizing enzymes of phosphoinositides such as PI and PI4,5P which help in mediating the Aβ assembly on lipid membranes affect the neuronal accumulation of Aβ, thereby impacting the AD pathogenesis (McLaurin and Chakrabartty 1997). The functionality of the PI4KIIIα complex, regulating the levels of plasmalemma PI4P and PI4,5P, remains highly conserved in Drosophila (Jeon et al. 2020). A recent study showed that the phenotypes of $A\beta_{1\text{-}42}$ expressing flies were mitigated through the genetic reduction of key components, including PI4KIIIα, Hyccin, tetratricopeptide repeat domain 7 (TTC7), and rolling blackout (RBO), within the PI4KIIIα complex (Zhang et al. 2017; Sun et al. 2018). This reduction led to a decrease in neuronal Aβ accumulation. Fourth, AD pathology on a systemic level is an energy-demanding process. The energy burden which eventually impairs systemic homeostasis occurs from several factors, such as imbalance in calcium influx, excessive pruning of synapses by the astrocytes and microglia, glutamate excitotoxicity and the toxic feedback loop created due to excessive glutamate concentration in the synapses. In addition to this, $A\beta_{1\text{-}42}$ causes oxidative stress. In the fly model of $A\beta_{1\text{-}42}$, genetic screening identified genes that were linked with the oxidative stress (Jeon et al. 2020). Endoplasmic stress (ER stress) has widely been observed in AD models, as well as in humans. Evidences from the fly model indicate that, in case of AD pathogenesis, ER stress response pathways might be strongly involved (Jeon et al. 2020).

To summarize, AD pathology entails a wide range of molecular disruptions that affect the cognitive and behavioral performance in humans. Research addressing the molecular parameters affecting the pathophysiology, and therapeutics in AD therefore needs to be investigated as per the requirement. The fly model given its advantage of a small lifespan, no ethical approval, easy handling, and genetic manipulation allows the experimenter to investigate several aspects of AD pathophysiology. Genetic screening and the impact of Aβ expression on a molecular level have been widely performed in the flies.

APPENDIX-1

Terminology

PD Parkinson's Disease
α-Syn α-Synuclein
FPD familial Parkinson's Disease
MPTP 1-methyl-4-phenyl-1,2,3,6-tetrahydropyridine
Poly Q polyglutamine
HD Huntington's disease
SCAs spinocerebellar ataxias

REFERENCES

Auluck, Pavan K, H Y Edwin Chan, John Q Trojanowski, Virginia M-Y Lee, and Nancy M Bonini. 2002. "Chaperone Suppression of α-Synuclein Toxicity in a Drosophila Model for Parkinson's Disease." *Science* 295 (5556): 865–68.

Bilen, Julide, and Nancy M Bonini. 2005. "Drosophila as a Model for Human Neurodegenerative Disease." *Annu. Rev. Genet.* 39: 153–71.

Carmine-Simmen, Katia, Thomas Proctor, Jakob Tschäpe, Burkhard Poeck, Tilman Triphan, Roland Strauss, and Doris Kretzschmar. 2009. "Neurotoxic Effects Induced by the Drosophila Amyloid-β Peptide Suggest a Conserved Toxic Function." *Neurobiology of Disease* 33 (2): 274–81.

Castellani, Rudy J, Germán Plascencia-Villa, and George Perry. 2019. "The Amyloid Cascade and Alzheimer's Disease Therapeutics: Theory versus Observation." *Laboratory Investigation* 99 (7): 958–70.

Cleary, James P, Dominic M Walsh, Jacki J Hofmeister, Ganesh M Shankar, Michael A Kuskowski, Dennis J Selkoe, and Karen H Ashe. 2005. "Natural Oligomers of the Amyloid-β Protein Specifically Disrupt Cognitive Function." *Nature Neuroscience* 8 (1): 79–84.

Crabtree, Donna M, and Jianhua Zhang. 2012. "Genetically Engineered Mouse Models of Parkinson's Disease." *Brain Research Bulletin* 88 (1): 13–32.

Crowther, D C, K J Kinghorn, E Miranda, R Page, J A Curry, F A I Duthie, D C Gubb, and D A Lomas. 2005. "Intraneuronal Aβ, Non-Amyloid Aggregates and Neurodegeneration in a Drosophila Model of Alzheimer's Disease." *Neuroscience* 132 (1): 123–35.

Feany, Mel B, and Welcome W Bender. 2000. "A Drosophila Model of Parkinson's Disease." *Nature* 404 (6776): 394–98.

Fernandez-Funez, Pedro, Maria Laura Nino-Rosales, Beatrice De Gouyon, Wei-Chi She, James M Luchak, Pedro Martinez, Enrique Turiegano, Jonathan Benito, Maria Capovilla, and Pamela J Skinner. 2000. "Identification of Genes That Modify Ataxin-1-Induced Neurodegeneration." *Nature* 408 (6808): 101–6.

Finelli, Alyce, Anju Kelkar, Ho-Juhn Song, Haidi Yang, and Mary Konsolaki. 2004. "A Model for Studying Alzheimer's Aβ42-Induced Toxicity in Drosophila Melanogaster." *Molecular and Cellular Neuroscience* 26 (3): 365–75.

Godena, Vinay K, Giulia Romano, Maurizio Romano, Chiara Appocher, Raffaella Klima, Emanuele Buratti, Francisco E Baralle, and Fabian Feiguin. 2011. "TDP-43 Regulates Drosophila Neuromuscular Junctions Growth by Modulating Futsch/MAP1B Levels and Synaptic Microtubules Organization." *PloS One* 6 (3): e17808.

Gong, Yuesong, Lei Chang, Kirsten L Viola, Pascale N Lacor, Mary P Lambert, Caleb E Finch, Grant A Krafft, and William L Klein. 2003. "Alzheimer's Disease-Affected Brain: Presence of Oligomeric Aβ Ligands (ADDLs) Suggests a Molecular Basis for Reversible Memory Loss." *Proceedings of the National Academy of Sciences* 100 (18): 10417–22.

Greeve, Isabell, Doris Kretzschmar, Jakob-Andreas Tschäpe, Anika Beyn, Claire Brellinger, Michaela Schweizer, Roger M Nitsch, and Rita Reifegerste. 2004. "Age-Dependent Neurodegeneration and Alzheimer-Amyloid Plaque Formation in Transgenic Drosophila." *Journal of Neuroscience* 24 (16): 3899–3906.

Hong, Soyon, Beth L Ostaszewski, Ting Yang, Tiernan T O'Malley, Ming Jin, Katsuhiko Yanagisawa, Shaomin Li, Tim Bartels, and Dennis J Selkoe. 2014. "Soluble Aβ Oligomers Are Rapidly Sequestered from Brain ISF in Vivo and Bind GM1 Ganglioside on Cellular Membranes." *Neuron* 82 (2): 308–19.

Iijima, Koichi, Hsueh-Cheng Chiang, Stephen A Hearn, Inessa Hakker, Anthony Gatt, Christopher Shenton, Linda Granger, Amy Leung, Kanae Iijima-Ando, and Yi Zhong. 2008. "Aβ42 Mutants with Different Aggregation Profiles Induce Distinct Pathologies in Drosophila." *PloS One* 3 (2): e1703.

Jackson, George R, Iris Salecker, Xinzhong Dong, Xiang Yao, Norman Arnheim, Peter W Faber, Marcy E MacDonald, and S Lawrence Zipursky. 1998. "Polyglutamine-Expanded Human Huntingtin Transgenes Induce Degeneration of Drosophila Photoreceptor Neurons." *Neuron* 21 (3): 633–42.

Jaiswal, Manoj Kumar, Wolf-Dieter Zech, Miriam Goos, Christine Leutbecher, Alberto Ferri, Annette Zippelius, Maria Teresa Carrì, Roland Nau, and Bernhard U Keller. 2009. "Impairment of Mitochondrial Calcium Handling in a MtSOD1 Cell Culture Model of Motoneuron Disease." *BMC Neuroscience* 10 (1): 1–16.

Jeon, Youngjae, Jae Ha Lee, Byoungyun Choi, So-Yoon Won, and Kyoung Sang Cho. 2020. "Genetic Dissection of Alzheimer's Disease Using Drosophila Models." *International Journal of Molecular Sciences* 21 (3): 884.

Knobloch, Marlen, Uwe Konietzko, Danielle C Krebs, and Roger M Nitsch. 2007. "Intracellular Abeta and Cognitive Deficits Precede Beta-Amyloid Deposition in Transgenic ArcAbeta Mice." *Neurobiology of Aging* 28 (9): 1297–1306. https://doi.org/10.1016/j.neurobiolaging.2006.06.019

Krüger, Rejko. 2004. "Genes in Familial Parkinsonism and Their Role in Sporadic Parkinson's Disease." *Journal of Neurology* 251: vi2–6.

Lambert, Mary P, A K Barlow, Brett A Chromy, C Edwards, R Freed, M Liosatos, T E Morgan, I Rozovsky, B Trommer, and Kirsten L Viola. 1998. "Diffusible, Nonfibrillar Ligands Derived from Aβ_{1-42} Are Potent Central Nervous System Neurotoxins." *Proceedings of the National Academy of Sciences* 95 (11): 6448–53.

Landles, C, K Sathasivam, A Weiss, B Woodman, H Moffitt, S Finkbeiner, B Sun, J Gafni, L M Ellerby, Y Trottier, and W G Richards. 2010. "Proteolysis of Mutant Huntingtin Produces an Exon 1 Fragment that Accumulates as an Aggregated Protein in Neuronal Nuclei in Huntington Disease." *Journal of Biological Chemistry* 285 (12): 8808–23.

Langston, J William, Philip Ballard, James W Tetrud, and Ian Irwin. 1983. "Chronic Parkinsonism in Humans Due to a Product of Meperidine-Analog Synthesis." *Science* 219 (4587): 979–80.

Lewis, Elizabeth A, and Gaynor A Smith. 2016. "Using Drosophila Models of Huntington's Disease as a Translatable Tool." *Journal of Neuroscience Methods* 265: 89–98.

Lu, Bingwei, and Hannes Vogel. 2009. "Drosophila Models of Neurodegenerative Diseases." *Annual Review of Pathology: Mechanisms of Disease* 4: 315–42.

Luheshi, Leila M, Gian Gaetano Tartaglia, Ann-Christin Brorsson, Amol P Pawar, Ian E Watson, Fabrizio Chiti, Michele Vendruscolo, David A Lomas, Christopher M Dobson, and Damian C Crowther. 2007. "Systematic in Vivo Analysis of the Intrinsic Determinants of Amyloid β Pathogenicity." *PLoS Biology* 5 (11): e290.

Luo, Liqun, Tim Tully, and Kalpana White. 1992. "Human Amyloid Precursor Protein Ameliorates Behavioral Deficit of Flies Deleted for Appl Gene." *Neuron* 9 (4): 595–605.

Martin, Ian, Valina L Dawson, and Ted M Dawson. 2011. "Recent Advances in the Genetics of Parkinson's Disease." *Annual Review of Genomics and Human Genetics* 12: 301–25.

McLaurin, JoAnne, and Avijit Chakrabartty. 1997. "Characterization of the Interactions of Alzheimer B-amyloid Peptides with Phospholipid Membranes." *European Journal of Biochemistry* 245 (2): 355–63.

Moloney, Aileen, David B Sattelle, David A Lomas, and Damian C Crowther. 2010. "Alzheimer's Disease: Insights from Drosophila Melanogaster Models." *Trends in Biochemical Sciences* 35 (4): 228–35.

Morris, Gary P, Ian A Clark, and Bryce Vissel. 2014. "Inconsistencies and Controversies Surrounding the Amyloid Hypothesis of Alzheimer's Disease." *Acta Neuropathologica Communications* 2: 1–21.

Mulakkal, Nitha C, Peter Nagy, Szabolcs Takats, Radu Tusco, Gábor Juhász, and Ioannis P Nezis. 2014. "Autophagy in Drosophila: From Historical Studies to Current Knowledge." *BioMed Research International* 2014: 273473.

Nixon, Ralph A, Dun-Sheng Yang, and Ju-Hyun Lee. 2008. "Neurodegenerative Lysosomal Disorders: A Continuum from Development to Late Age." *Autophagy* 4 (5): 590–99.

Pandey, Udai Bhan, and Charles D Nichols. 2011. "Human Disease Models in Drosophila Melanogaster and the Role of the Fly in Therapeutic Drug Discovery." *Pharmacological Reviews* 63 (2): 411–36.

Pandey, Udai Bhan, Zhiping Nie, Yakup Batlevi, Brett A McCray, Gillian P Ritson, Natalia B Nedelsky, Stephanie L Schwartz, Nicholas A DiProspero, Melanie A Knight, and Oren Schuldiner. 2007. "HDAC6 Rescues Neurodegeneration and Provides an Essential Link between Autophagy and the UPS." *Nature* 447 (7146): 860–64.

Pawar, Amol P, Kateri F Dubay, Jesus Zurdo, Fabrizio Chiti, Michele Vendruscolo, and Christopher M Dobson. 2005. "Prediction of 'Aggregation-Prone' and 'Aggregation-Susceptible' Regions in Proteins Associated with Neurodegenerative Diseases." *Journal of Molecular Biology* 350 (2): 379–92.

Periz, Goran, and Mark E Fortini. 2004. "Functional Reconstitution of Γ-secretase through Coordinated Expression of Presenilin, Nicastrin, Aph-1, and Pen-2." *Journal of Neuroscience Research* 77 (3): 309–22.

Pimplikar, Sanjay W. 2009. "Reassessing the Amyloid Cascade Hypothesis of Alzheimer's Disease." *The International Journal of Biochemistry & Cell Biology* 41 (6): 1261–68.

Prakash, Pavitra, Arpit Kumar Pradhan, and Vasu Sheeba. 2022. "Hsp40 Overexpression in Pacemaker Neurons Delays Circadian Dysfunction in a Drosophila Model of Huntington's Disease." *Disease Models & Mechanisms* 15 (6): dmm049447.

Prüßing, Katja, Aaron Voigt, and Jörg B Schulz. 2013. "Drosophila Melanogaster as a Model Organism for Alzheimer's Disease." *Molecular Neurodegeneration* 8 (1): 1–12.

Ratnaparkhi, Anuradha, George M Lawless, Felix E Schweizer, Peyman Golshani, and George R Jackson. 2008. "A Drosophila Model of ALS: Human ALS-Associated Mutation in VAP33A Suggests a Dominant Negative Mechanism." *PloS One* 3 (6): e2334.

Reitz, Christiane. 2012. "Alzheimer's Disease and the Amyloid Cascade Hypothesis: A Critical Review." *International Journal of Alzheimer's Disease* 2012: 369808.

Schapira, A H V, J M Cooper, D Dexter, J B Clark, P Jenner, and C D Marsden. 1990. "Mitochondrial Complex I Deficiency in Parkinson's Disease." *Journal of Neurochemistry* 54 (3): 823–27.

Simonsen, Anne, Robert C Cumming, Andreas Brech, Pauline Isakson, David R Schubert, and Kim D Finley. 2008. "Promoting Basal Levels of Autophagy in the Nervous System Enhances Longevity and Oxidant Resistance in Adult Drosophila." *Autophagy* 4 (2): 176–84.

Spillantini, Maria Grazia, Marie Luise Schmidt, Virginia M-Y Lee, John Q Trojanowski, Ross Jakes, and Michel Goedert. 1997. "α-Synuclein in Lewy Bodies." *Nature* 388 (6645): 839–40.

Sun, Minghao, Yinghui Zhao, Men Han, Baozhu Zhang, Xiao Zhang, Qichao Zhang, Nastasia K-H Lim, Wen-An Wang, and Fu-De Huang. 2018. "TTC7 and Hyccin Regulate Neuronal Aβ 42 Accumulation and Its Associated Neural Deficits in Aβ 42-Expressing Drosophila." *Journal of Alzheimer's Disease* 65 (3): 1001–10.

Townsend, Matthew, Ganesh M Shankar, Tapan Mehta, Dominic M Walsh, and Dennis J Selkoe. 2006. "Effects of Secreted Oligomers of Amyloid B-protein on Hippocampal Synaptic Plasticity: A Potent Role for Trimers." *The Journal of Physiology* 572 (2): 477–92.

Walsh, Dominic M, Igor Klyubin, Julia V Fadeeva, William K Cullen, Roger Anwyl, Michael S Wolfe, Michael J Rowan, and Dennis J Selkoe. 2002. "Naturally Secreted Oligomers of Amyloid β Protein Potently Inhibit Hippocampal Long-Term Potentiation in Vivo." *Nature* 416 (6880): 535–39.

Wang, C E, S Tydlacka, A L Orr, et al. 2008. "Accumulation of N-terminal Mutant Huntingtin in Mouse and Monkey Models Implicated as a Pathogenic Mechanism in Huntington's Disease." *Human Molecular Genetics* 17 (17): 2738–51.

Warrick, John M, H Y Chan, Gladys L Gray-Board, Yaohui Chai, Henry L Paulson, and Nancy M Bonini. 1999. "Suppression of Polyglutamine-Mediated Neurodegeneration in Drosophila by the Molecular Chaperone HSP70." *Nature Genetics* 23 (4): 425–28.

Ye, Yihong, and Mark E Fortini. 1999. "Apoptotic Activities of Wild-Type and Alzheimer's Disease-Related Mutant Presenilins in Drosophila Melanogaster." *The Journal of Cell Biology* 146 (6): 1351–64.

Zhang, Xiao, Wen-An Wang, Li-Xiang Jiang, Hai-Yan Liu, Bao-Zhu Zhang, Nastasia Lim, Qing-Yi Li, and Fu-De Huang. 2017. "Downregulation of RBO-PI4KIIIα Facilitates Aβ42 Secretion and Ameliorates Neural Deficits in Aβ42-Expressing Drosophila." *Journal of Neuroscience* 37 (19): 4928–41.

Index

S

T

V